Controlled Release Technologies: Methods, Theory, and Applications

Volume II

Editor

Agis F. Kydonieus, Ph.D.
Executive Vice President
Hercon Group
Health-Chem Corporation
New York, New York

CRC Press
Taylor & Francis Group
Boca Raton London New York

CRC Press is an imprint of the
Taylor & Francis Group, an **informa** business

CRC Press
Taylor & Francis Group
6000 Broken Sound Parkway NW, Suite 300
Boca Raton, FL 33487-2742

Reissued 2019 by CRC Press

A Library of Congress record exists under LC control number:

Publisher's Note
The publisher has gone to great lengths to ensure the quality of this reprint but points out that some imperfections in the original copies may be apparent.

Disclaimer
The publisher has made every effort to trace copyright holders and welcomes correspondence from those they have been unable to contact.

ISBN 13: 978-0-367-25370-7 (hbk)
ISBN 13: 978-0-367-25371-4 (pbk)
ISBN 13: 978-0-429-28742-8 (ebk)

Visit the Taylor & Francis Web site at http://www.taylorandfrancis.com and the
CRC Press Web site at http://www.crcpress.com

PREFACE

Delivery of chemical ingredients by controlled release processes occurs in nature. Examples of such processes include the delivery and control of the flow of food and waste across the exterior membrane of one-celled animals and the oxygenation of blood in mammals by the diffusion of oxygen through the alveolar walls. Development of controlled release systems for our modern technology may be considered an attempt to simulate nature's processes and thereby provide more efficient and more effective delivery of chemicals to intended targets. Information and knowledge gained in these exploits form the substance of the two volumes of this treatise.

Intensive research and development by the drug industry in the last 20 years has provided the scientific foundation for this fledging new science. Concern over the administration of high doses of conventional drugs was the impetus toward the development of controlled release oral drugs initially and implantable devices more recently. Applying much the same principles, scientists working in the pesticide field have developed new technologies and formulations to (1) extend the duration of effectiveness of pesticides without increasing rates of application, (2) reduce the hazard associated with the use of highly poisonous chemicals, (3) prolong the effective life of unstable, volatile, or hydrolyzable pesticides, (4) improve pest control efficiency, and (5) minimize pollution of the environment.

The technologies described in these two volumes depend almost exclusively on the use of polymers and polymer technology. Exceptional advancement in polymer science during the past several decades has made possible the creation of delivery systems which previously could not have been conceived. Described herein are dozens of polymers that comprise the elements of delivery systems that control the permeation and release of active ingredients.

The applications, advantages, and fundamental concepts of controlled release have been the subject of many symposia and of several books dealing with formulations: however, all known delivery systems have not been assembled for consideration as they are here.

Our emphasis in these volumes is on the description of the controlled release technologies, their theoretical basis, mechanisms of release, and their advantages and limitations. Many commercial applications, as well as experimental products and formulations, are discussed to illustrate the potential of each of these technologies. Young as this technology may be, many scientists are convinced that it has the momentum and the secure fundamental basis for solving the complex problems of the delivery of drugs to humans and of directing pesticides to targeted biota with minimum disruption of environmental values. Formulations have already been developed that have been proven commercially successful in many fields, including medicine, agriculture, forestry, public health, and products for the consumer and industry. At this juncture, our aim is that this volume will add to the current momentum to explore this already proven technology and that it will help product development scientists choose the proper controlled release technology for their specific product needs.

The fundamental concepts and theoretical background of controlled release processes are given in Volume 1, Chapter 1. Having introduced the subject matter in the introductory chapter, Chapters 2 through 6 and Volume II, Chapters 1 through 12 follow, describing in detail 17 different controlled release technologies; Chapter 13 describes briefly several other controlled release technologies and provides a 1977 to 1978 patent search of controlled release methods and applications. Each chapter was written by an author who either invented a specific controlled release technology or has had a major roll in advancing the state of its art. Representing all the major technological areas of controlled release, the subject matter is grouped as follows in the chapters indicated:

1. Volume I, Chapters 2 through 6: monolithic devices, membranes, porous and homogenous films, and laminated structures.
2. Volume II, Chapters 1 and 2: erodible, bioerodible, soluble, hydrolyzable, and biodegradable devices.
3. Chapter 3: retrograde chemical reaction systems.
4. Chapters 4 through 8: microencapsulation systems by coascervation, interfacial polymerization, air suspension, and centrifugal extrusion.
5. Chapters 9 through 12: other important controlled release technologies, such as hollow fibers, osmotic devices, starch xanthate matrices, and kraft lignin carriers.

As an introductory background and perspective for the novice or the recent practitioner in the field, the first chapter was included to introduce all the major controlled release technologies and to describe the basic components of controlled release devices and the release characteristics of the controlled release processes. New developments and improvements in existing technologies continue to appear. Described in Chapter 13 are some new and improved technologies and products, actually those found in a review of the Official Gazettes of the U.S. Patent and Trademark Office for 1977 and part of 1978. Several other controlled release technologies which were found in the literature have also been included in Chapter 13. Obviously, in the fast-expanding field of controlled release, several omissions must have occured despite our effort to present all significant developments known by the end of 1977, when most of the manuscripts were collected. Nevertheless, we hope that this effort will prove to be of value to scientists and product development engineers seeking up-to-date information in the field.

Several friends and associates should be given credit for their helpful suggestions and criticisms. Special thanks should go to Dr. Morton Beroza for scrutinizing parts of the manuscript and to Drs. Nate Cardarelli and Bill O'Neill for their guidance and encouragement. I am also indebted to the authors for their cooperation in adhering to strict manuscript specifications and to Ms. Camille Boxhill and Ms. Adriane Chisholm for their efforts in typing and assisting in the editorial endeavors. Finally, I would like to thank the management of Hercon Group and Health-Chem Corporation, who have been strong advocates of controlled release for many years and have given the editor all the support required to complete this undertaking.

Agis F. Kydonieus
September 1978

THE EDITOR

Agis Kydonieus, Ph.D., is Executive Vice President of Hercon Products Group, a division of the Health-Chem Corporation, New York, N.Y. and Adjunct Professor of Chemical Engineering, The Cooper Union, New York, N.Y.

Dr. Kydonieus graduated in 1959 from the University of Florida, Gainesville, Florida, with a BS degree in Chemical Engineering (summa cum laude) and obtained his Ph.D. degree from the same school in 1964.

Dr. Kydonieus is a member of the Board of Governors of The Controlled Release Society as well as a member of The American Institute of Chemical Engineers, The Entomological Society of America, American Chemical Society, Pest Control Association, and The Society of Plastics Engineers.

Dr. Kydonieus is the author of several dozen patents, publications and presentations in the fields of controlled release and biomedical devices.

CONTRIBUTORS

G. Graham Allan, Ph.D., D.Sc.
Professor of Fiber and Polymer
  Science
Department of Chemical Engineering
College of Forest Resources
University of Washington
Seattle, Washington

Joseph A. Bakan, B.S.
Director of Research
Capsular Products Division
Appleton Papers, Inc.
Dayton, Ohio

John W. Beer
Research Associate
Centro Agronomico Tropical de
Investigaciony Ensenanza
Turriabla, Costa Rica

Thomas W. Brooks, Ph.D.
Technical Director
Conrel
Needham Heights, Massachusetts

Massimo Calanchi
Manager
Research and Development
Microcaps
Eurand, S.p.A.
Milan, Italy

Nate F. Cardarelli, M.S.
Director
Environmental Management
  Laboratory
University of Akron
Akron, Ohio

Michael J. Cousin, Ph.D.
Research Scientist
Battelle Columbus Laboratories
Columbus, Ohio

H. T. DelliColli, Ph.D.
Product Development Manager
Agrichemical Products
Westvaco Corporation
North Charleston, South Carolina

John T. Goodwin, Ph.D.
Vice President
Chemistry and Chemical Engineering
  Division
Southwest Research Institute
San Antonio, Texas

Harlan S. Hall, B.S.
President
Coating Place, Inc.
Verona, Wisconsin

Frank W. Harris, Ph.D.
Professor of Chemistry
Wright State University
Dayton, Ohio

Robert C. Koestler, Ph.D.
Project Leader—Microencapsulation
Pennwalt Corporation
King of Prussia, Pennsylvania

Mario Maccari, Ph.D.
Director
Research and Development
Eurand S.p.A.
Milan, Italy

Ruth Ann Mikels, B.S.
Research Assistant
College of Forest Resources
University of Washington
Seattle, Washington

Arthur S. Obermayer, Ph.D.
President
Moleculon Research Corporation
Cambridge, Massachusetts

William P. O'Neill, Ph.D., M.B.A.
Management and Technical
  Consultant
Poly-Planning Services
Los Altos, California

Ralph E. Pondell, B.S.
Vice President-Secretary
Coating Place, Inc.
Verona, Wisconsin

Alberto R. Quisumbing, Ph.D
Director Field Applications
Hercon Group
Herculite Products, Inc.
New York, New York

Theodore J. Roseman, Ph.D.
Senior Research Scientist
The Upjohn Company
Kalamazoo, Michigan

Mario F. Sartori, Ph.D.
Consultant
Department of Chemistry
University of Delaware
Newark, Delaware

Baruch Shasha, Ph.D.
Research Chemist
Northern Regional Research Center
Agricultural Research
Science and Education Administration
United States Department of
  Agriculture
Peoria, Illinois

George R. Somerville, B.S.
Director
Department of Applied Chemistry and
  Chemical Engineering
Southwest Research Institute
San Antonio, Texas

Felix T. Theeuwes, Sc.D.
Program Director
OROS Products
Principal Scientist
Alza Research
Palo Alto, California

Seymour Yolles, Ph.D.
Professor of Chemistry
University of Delaware
Newark, Delaware

TABLE OF CONTENTS

Volume I

Volume II

Chapter 1

ERODIBLE MATRICES

S. Yolles and M. F. Sartori

TABLE OF CONTENTS

I. INTRODUCTION

The system most investigated for controlled release comprises incorporating a biologically active agent in a matrix, commonly a polymeric material, shaping the obtained composite into a convenient form, and implanting it into the body tissue or body cavity by surgery or injection.

Early investigations on controlled release delivery systems have used polyethylene film,[1] silicone rubber,[2,3] etc., as matrices. They are effective matrices, but with the disadvantage that, after the implanted composites have fulfilled their functions, the matrices require surgical removal.

Composites of drugs with erodible matrices, which are consumed or biodegraded during therapy, were developed by Yolles[4,5,6] in the study of controlled release of narcotic antagonists.

Specific requirements for a polymeric material to be used as a matrix are (1) compatibility with the environment and with the biologically active ingredient, (2) rate of biodegradation compatible with the projected life span of the composite, (3) ability to provide the desired release rate, and (4) availability of starting material.

Erodible polymers and copolymers investigated most are (1) poly(lactic acid), (2) poly(glycolic acid), (3) copolymers of poly(lactic acid) and poly(glycolic acid), and (4) polyamides and copolymers of polyamides and polyesters. Their uses, preparation and properties are described below.

Several methods have been used to prepare composites containing erodible matrices: dissolving the polymer and the active ingredient with[7] or without plasticizer[8] in a solvent, evaporating the solvent and press-melting the residue between plates in a heated press to produce films; grinding the polymer, mixing with the active ingredient, and compression molding or extruding the mixture into films or pellets[7,9] (pelletizing has been conveniently accomplished with a Marumerizer®[10]); covalently bonding the drug with reactive groups of the polymer and preparing films as above.[11] Films have been reformed into beads, chips, etc. Tubes[12] and rods[9] also have been made.

II. POLY(LACTIC ACID)

Poly(lactic acid) (PLA), sometimes referred to as polylactide, was reported by Kulkarni[13] in 1966 to be a suitable material for surgical implant because it undergoes hydrolytic de-esterification to lactic acid, a normal product of muscle metabolism. At about the same time and independently, Schneider[14] disclosed that PLA was useful for preparing absorbable sutures. The first use of PLA as an erodible matrix for controlled release of drugs was reported by Yolles in 1971.[5] Since then many patents[15,16] and papers[7,12,17,18] have appeared concerning the preparation and the uses of this polymer.

Since lactic acid is optically active, PLA can exist in the following forms: RS, −R and +S forms. Both the +S and RS-PLA have been found to be free of harmful tissue reaction upon implantation.[8]

A. Uses

The most widely reported uses of PLA as a matrix for controlled release is in systems containing the following compounds: (1) narcotic antagonists,[7,21,22] (2) fertility control agents,[12,23] (3) anticancer agents,[18] and (4) herbicides and pesticides.[24]

B. Preparation

Dilactide, the cyclic diester of lactic acid, is polymerized into PLA by heating at 130°C[7] or 170°C[13] under vacuum[12] or at atmospheric pressure[7] and in the presence of nucleophiles, such as tertiary amines,[12] or of suitable catalysts, such as stannous octoate,[9] stannous chloride,[13] $ZnCl_2$,[13] PbO,[13] SnO,[13] ZnO,[13] tetraphenyl tin,[13] or diethyl zinc.[7] The last two catalysts are preferred. The white fine powder which forms is purified by precipitation from dichloromethane.

C. Properties

Poly(S-lactic acid) is highly crystalline with a mp $T_m = 180°C$ and a glass transition temperature around 67°C. Poly(RS-lactic acid) is amorphous and has a glass transition temperature around 57°C.[12] Poly(RS-lactic acid) is soluble in most common organic solvents, such as benzene, acetone, tetrahydrofuran, and chlorinated hydrocarbons. Poly(S-lactic acid) is soluble in halocarbons, e.g., chloroform and methylene chloride at 50 to 60°C.[12]

The molecular weight of PLA can be controlled by changing the nature or quantity of the catalyst.[13] With tin catalysts PLA can be obtained with $\bar{M}_n$ up to about 400,000 ($[\eta]$ in chloroform about 7.3 dℓ/g).[20] With diethyl zinc, samples of PLA were prepared with mol wt of 45,000 and 70,000 (calculated from viscosimetric measurements).[7] The intrinsic viscosity of PLA, prepared as in Reference 13, in benzene (100 mℓ/g) varies from 0.5 to 1.18.[13]

The shelf life of poly(lactic acid) is good, although a decrease in molecular weight with time is observed depending on purity of the polymer.[12] Samples prepared by using -S-lactide and diethyl zinc as the catalyst show hydrolytic stability. PLA degrades in vivo, losing 12 to 14% in 3 months, according to Kulkarni,[13] whereas in vivo and in vitro experiments by Schindler[12] show that it takes about 80 days for a PLA composite to halve its average molecular weight. The authors of this paper have found that the life of the polymer is proportional to molecular weight, the higher the $\bar{M}_n$, the slower the erosion in vivo. Carboxyl end groups are given by the equation

$$[COOH] = 1.77 \times 10^{-5} \ (\Delta E/C_p)$$

where ΔE is the change in optical density at 515 nm and C_P is polymer concentration in g/1.[12] Values of COOH group content, reported in the literature, vary between 2 and 10 mg KOH per gram polymer.[7] Such end group analyses are useful in determining $\bar{M}_n$.

PLA is sensitive to γ-radiation.[12] Histological studies indicate that PLA is nontoxic, nontissue reactive, and biodegradable. The degradation products are not retained in any of the vital organs of animals.[13]

III. POLY(GLYCOLIC ACID)

The preparation of poly(glycolic acid) (PGA) was disclosed by Higgins in 1954.[25] Most of the uses of this polymer are in the preparation of systems for sutures in surgery,[16,26,28] in prosthetic devices,[29] and in storage pellets.[30] Dexon® is the trade name for PGA used as sutures.

A. Preparation

One preferred route to prepare PGA involves the polymerization of glycolide, the cyclic dimeric condensation product obtained by dehydrating glycolic acid. The polymerization occurs by heating at 155 to 241°C in the presence of antimony trioxide[31] or at 150°C in the presence of amines.[32]

B. Properties

The molecular weights of PGA can be controlled by varying the preparation conditions. Preferably the molecular weight is in the range of 10,000 or more.[29] The straight-pull tensile strength is greater than that of catgut and silk of comparable diameter.[33]

Absorption studies in animals show that complete absorption usually occurs within 90 days and no pathologic changes are detectable in various tissues examined in 9 months.[34] It is completely nontoxic upon implantation into animals.[34]

Because of its biodegradability and nontoxic properties, PGA should be a good candidate as a matrix for controlled release. However, its use is limited because of difficulties in fabricating composites, e.g., its low solubility in common solvents.

IV. POLY(LACTIC ACID)/POLY(GLYCOLIC ACID)

The first use of this copolymer (PLA/PGA) as an erodible matrix was reported in 1967.[35] Since then various patents[36,37,38] and papers[39,40,41,42] have been published concerning the preparation of these copolymers and their use for the controlled release of drugs, fertilizers, and insecticides.

A. Preparation
Lactide and glycolide are polymerized into PLA/PGA copolymers by heating at temperatures varying from 115 to 135°C under vacuum and in the presence of stannous octoate as the catalyst.[38,39]

Copolymers of lactic acid (LA) and glycolic acid (GA) ranging from 75 LA/25 GA,[9] 90 LA/10 GA[5,9] and 50/50 RS/lactide/glycolide[39] have been prepared.

B. Properties
Copolymer 90/10 is an opaque, snow-white crystalline material, whereas the 50/50 copolymer is an orange transparent material.[39] The temperature range at which these copolymers become rubbery is 55 to 60°C.[5] The molecular weights determined by gel permeation chromatography, membrane osmosis, and light scattering, range from 40,000 to 200,000.[8] Intrinsic viscosities measured at 37°C in tetrahydrofuran, range from 0.4 to 1.0.[8]

V. POLYAMIDES AND POLYAMIDE-POLYESTER COPOLYMERS

Numerous articles and patents are reported in the literature describing properties and uses of polyamides and of copolymers of polyamides and polyesters.[43,44,45,46,47] However, very little is reported on the erodibility of these polymers.[48] Mixtures of polyamides and ethylcellulose have been used as suture material.[49] Collagen polymers are reported to be completely resorbed without toxic actions.[50] Sidman, et al.[51] investigated glutamic acid/leucine copolymers in the preparation of biodegradable delivery systems for narcotic antagonists. The composition of the copolymers varied from 10% glutamic acid/90% leucine to 40% glutamic acid/60% leucine. The preparation of composites with naltrexone in the forms of films, rods, and tubes is described.[51]

An erodible intrauterine device has been made by crosslinking gelatin with ketones and incorporating prostaglandin.[52]

VI. MISCELLANEOUS MATRICES

In searching the literature for erodible matrices, very few references have been found on the following polymers and copolymers and their use as matrices: polyester from succinic acid and ethylene glycol;[53] poly (ε-caprolactone);[54] copolymer of dilactide and ε-decalactone[54] and copolymer of ε-decalactone and ε-caprolactone.[54] However, investigation in this field of controlled release is going forward at a feverish rate, including the search for new and patentable bioerodible polymers.

Recently one of the authors[11] of this paper has developed an erodible system in which the active ingredient is covalently bonded (ester linkage) to water-soluble Klucel®[55] (hydroxypropyl cellulose). The compound is water-insoluble. The release mechanism occurs by hydrolysis of the ester linkage (enzymatic in vivo), followed by diffusion of the active ingredient from the polymer. After the drug has been delivered, the polymer matrix becomes water-soluble again and disappears from the injection site. The same principle is now being investigated using an amide bond between drug and water-soluble cellulosic.

REFERENCES

1. **Yolles, S.,** Development of a Long-Acting Dosage Form for Narcotic Antagonists, paper presented at the 13th Nat. Med. Chem. Symp., University of Iowa, Iowa City, June 18 to 22, 1972.
2. **Doyle, L. L. and Clewe, T. H.,** Preliminary studies on the effect of hormone releasing intrauterine devices, *Am. J. Obstet. Gynecol.*, 101, 564, 1968.
3. **Roseman, T. J. and Higuchi, W. I.,** Release of medoxyprogestrone acetate from a silicon polymer, *J. Pharm. Sci.*, 59, 353, 1970.
4. **Yolles, S., Eldridge, J. E., and Woodland, J. H. R.,** Sustained delivery of drugs from polymer/drug mixtures, *Polym. News*, 1, 9, 1971.
5. **Blake, D. A., Yolles, S., Helrich, M., Cascorbi, H. F., and Eagan, M. J.,** Release of Cyclazocine from Subcutaneously Implanted Polymeric Matrices, Abstract, Academy of Pharmaceutical Sciences, San Francisco, March 30, 1971.
6. **Willette, R. E.,** The search for long-acting preparations, in *Narcotic Antagonists*, Willette, R. E., Ed., The National Institute of Drug Abuse, Rockville, Md., 1976, 1.
7. **Woodland, J. H. R., Yolles, S., Blake, D., Helrich, M., and Meyer, F. J.,** Long-acting delivery systems for narcotic antagonists, *J. Med. Chem.*, 16, 899, 1973.
8. **Schwope, A. D., Wise, D. L., and Howes, J. F.,** Lactic/glycolic acid polymers as narcotic antagonist delivery systems, in *Life Sciences*, Vol. 17, Pergamon Press, New York, 1975, 1878.
9. **Sinclair, R. G. and Gynn, G. M.,** Preparation and evaluation of glycolic and lactic acid-based polymers, in *Final Scientific Report*, Part II, Contract No. DADA 17-72-2066, U.S. Army Institute of Dental Research, Walter Reed Army Medical Center, Washington, D.C., 1972, 6.
10. Elanco Product Co., Division of Eli Lilly & Co., Indianapolis, Indiana.
11. **Yolles, S.,** Time release depot for anticancer drugs: release of drugs covalently bonded to polymers, *J. Parenter. Drug Assoc.*, 32, 188, 1978.
12. **Schindler, A., Coat, J. F., Kimble, G., Pitt, C. G., Wall, M. E., and Zweidinger, R.,** Biodegradable polymers for sustained drug delivery, in *Contemporary Topics in Polymer Science*, Vol. 2, Pearce, E. M. and Schaefgen, J. R., Eds., Plenum Press, New York, 1977, 264.
13. **Kulkarni, R. K., Pani, K., Neuman, C., and Leonard, F.,** Poly(lactic acid) for surgical implants. Technical Rep. 6608, Walter Reed Army Medical Center, Washington, D.C., 1966.
14. **Schneider, A. K.,** French Patent 1,478,694, 1967.
15. **Ramsey, W. B. and DeLapp, D. F.,** U.S. Patent 3,781,349, 1973.
16. **Yolles, S.,** U.S. Patent 3,887,699, 1975.
17. **Jackanicz, T. M., Nash, H. H., Wise, D. L., and Gregory, J. B.,** Poly(lactic acid) as a biodegradable carrier for contraceptive steroids, *Contraception*, 8, 227, 1973.
18. **Yolles, S., Leafe, T. D., and Meyer, J.,** Controlled release of anticancer agents, *J. Pharm. Sci.*, 64, 115, 1975.
19. **Boswell, G. A. and Scribner, R. M.,** U. S. Patent 3,773,919, 1973.
20. **Dittrich, W. and Schulz, R. C.,** Kinetics and mechanisms of the ring-opening polymerization of lactide, *Angew. Makromol. Chem.*, 15, 109, 1971.
21. **Leafe, T. D., Sarner, S., Woodland, J. H. R., Yolles, S., Blake, D. H., and Meyer, F. J.,** Narcotic antagonists, in *Advances in Biochemical Psycopharmacology*, Vol. 8, Brand, M. C., Harris, L. S., May, E. L., Smith, J. P., and Villareal, J. E., Eds., Raven Press, New York, 1973, 569.
22. **Yolles, S., Leafe, T. D., Woodland, J. H. R., and Meyer, F. J.,** Release rates of naltrexone from PLA, *J. Pharm. Sci.*, 64, 348, 1975.

23. **Yolles, S., Leafe, T. D., Sartori, M. S., Torkelson, M., and Ward, L.,** Controlled release of biologically active agents, in *Controlled Release Polymeric Formulations*, ACS Symp. Ser. No. 33, Paul, D. R. and Harris, F. W., Eds., American Chemical Society, Washington, D.C., 1976, 125.

24. **Sinclair, R. G.,** Slow release pesticide system, *Environ. Sci. and Technol.*, 7, 955, 1973.

25. **Higgins, W. A.,** U.S. Patent 2,676,945, 1954.

26. **Furgivele, F. P.,** Ophthalmic use of "Dexon", *Ann. Ophthalmol.*, 6, 1219, 1974.

27. **Aird, C. C. and Matory, W. F.,** Evaluation of "Dexon" suture in dogs, *J. Nat. Med. Assoc.*, 66, 424, 1974.

28. **Kuo, T. P.,** Clinical use of PGA in vascular surgery, *J. Formosan Med. Assoc.*, 73, 45, 1974.

29. **Schmitt, E. E.,** U.S. Patents 3,463,158, 1969; 3,739,773, 1973.

30. **Schmitt, E. E. and Polistina, R. A.,** U.S. Patent 3,991,766, 1976.

31. **Lowe, C. E.,** U.S. Patent 2,668,162, 1954.

32. Shell International, Belgium Patent 657,429, 1965.

33. Davis and Geck Co., Manufacturer's Data, Danbury, Conn.

34. **Darkik, H.,** Clinical use of PGA as a new absorbable synthetic suture, *Am. J. Surg.*, 121, 656, 1971.

35. **Schmitt, E. E. and Polistina, R. A.,** U.S. Patent 3,297,033, 1967.

36. **Schmitt, E. E.,** South African Patent 71-08.150, 1972.

37. **Glick, A.,** U.S. Patent 3,772,420, 1973.

38. **Wasserman, D.,** U.S. Patent 3,839,297, 1974.

39. **Sinclair, R. G. and Gynn, G. M.,** U.S. Army Medical Research and Development Command Report, Rm 8H089, Forrestal Building, Washington, D.C., February 16, 1972.

40. **Hegyeli, A. F.,** Biodegradation of PLA/PGA implant materials, *J. Biomed. Mater. Res.*, 7, 205, 1973.

41. **Cutright, D. E.,** Degradation of polymers and copolymers of PLA/PGA, *Oral Surg.*, 37, 142, 1974.

42. **Sinclair, R. B.,** Slow release pesticide system, *Environ. Sci. Technol.*, 7, 956, 1973.

43. **Kurtz, L.D.,** French Patent 2,059,690, 1971.

44. **Kurtz, L. D.,** French Patent 2,059,691, 1971.

45. **Mori, S. and Akamatsu, A.,** Ajiromoto Co., Inc., Japanese Patent 72-43,220, 1972.

46. **Sciarra, J. J. and Gidwani, R. N.,** Polyamide with drug, *J. Pharm. Sci.*, 61, 754, 1972.

47. **Mori, S., Akamatsu, A., and Togo, K.,** German Patent 2,218,200, 1972.

48. **Kurtz, L. D.,** U.S. Patent 3,642,003, 1970.

49. **Sciarra, J. J. and Gidwani, R. N.,** Mixtures of ethylcellulose and polyamides as suture materials, *Soc. Cosmet. Chem.*, 21, 667, 1970.

50. **Oleinik, E. M.,** Collagen polymers used in surgery, *Gig. Vop. Proizvod. Primen. Polim. Mater. (Russian)* 249, 1969.

51. **Sidman, K. R., Arnold, D. L., Steber, W. D., Nelsen, L., Granchelli, F. E., Strong, P., and Sheth, S. G.,** Use of synthetic polypeptides in the preparation of biodegradable delivery vehicles for narcotic antagonists, in *Narcotic Antagonists*, Willette, R., Ed., The National Institute of Drug Abuse, Rockville, Md., 1976, 33.

52. **Ramwell, P. W.,** U.S. Patent 3,888,975, 1975.

53. **Coquard, J., Sedivy, P., Verrier, J., and Ruand, M.,** U.S. Patent 3,883,901, 1975.

54. **Schindler, A., Coat, J. F., Kimble, G., Pitt, C. G., Wall, M. E., and Zweidinger, R.,** Biodegradable polymers for sustained drug delivery, in *Contemporary Topics in Polymer Science*, Vol. 2, Pearce, E. M. and Schaefgen, J. R., Eds., Plenum Press, New York, 1977, 271.

55. Klucel®, Pamphlet Published by Hercules, Inc., Wilmington, Del., 1971, 4.

Chapter 2

THE BIODEGRADATIVE CONTROLLED RELEASE OF PESTICIDES FROM POLYMERIC SUBSTRATES

G. G. Allan, J. W. Beer, M. J. Cousin, and R. A. Mikels

TABLE OF CONTENTS

I. GENERAL DESCRIPTION

The controlled release of pesticides from biodegradable substrates is not a topic new to Nature. It is only new to contemporary man. Many of the higher plants have evolved systems for controlling contiguous competitive vegetation by a phenomenon known as allelopathy.[1] For instance, the absence of weeds around the base of the tree, *Eucalyptus camaldulensis,* is attributed to herbicidal terpenoid compounds produced in the leaves and delivered to the soil surface via rain water runoff.[2] Similarly, the biological decay of the stalks and dead roots of some grasses and shrubs liberates materials that are phytotoxic to vicinal vegetative species.[3] Thus, these compounds are beneficial to the existing and succeeding generations of the mother plant. In many of these cases, the natural controlled release mechanisms provide nonactive glycosides of the phytotoxicant.[4] Therefore, further hydrolytic degradation must take place before the ordained herbicidal effect is observed. This exemplifies Nature's way of utilizing biodegradable controlled release systems to maintain phytotoxic chemical concentrations. In this connection, it is worthy of note that the widespread existence of allelopathic systems, as well as the common occurrence of zootoxic plants, should dispel the myth that Nature's products, especially after they have been biodegraded, are innately harmless.

Independent of these naturally evolved systems, many different formulation methods for controlling the release of herbicides have been envisioned and investigated[5] and these are described elsewhere in this volume. This chapter will focus on the release of chemically attached herbicides from biodegradable substrates as exemplified by Figure 1, although obviously the system has applicability to biocides in general.

II. THE CONCEPT OF BIODEGRADABILITY

First, however, the concept of biodegradability must be set forth. This is usually defined as a time-dependent decomposition process originating with the stresses induced by the chemical and biochemical action of the environment. A contemporary rider frequently attached to this definition states that these biodegradable substances will preferably be natural products. Of course, within this definition several deficiencies are immediately obvious. The most serious of these is the vagueness of the specification of the time required for a material to disintegrate and thereby earn the label "biodegradable". In other words, it is important to realize that, given enough time, all things (including even mountains) will eventually entropically degrade. Clearly, some reasonable limit for the decomposition time must be agreed upon so that a systematic classification of biodegradable materials can be drawn up. Nonetheless, such an agreement is likely to prove elusive. The general aversion to plastic packaging litter on beaches and roadsides, for example, is testimony that the breakdown of some synthetic polymers is not sufficiently rapid to be broadly acceptable. These slow rates of biodegradation are somewhat surprising because soil microorganisms show an exceptional capability of adapting to new carbon nutrient sources.[6] The crucial rate determining factor is probably the equilibrium moisture content of the plastic. If this is very low, microorganisms will have no vehicle to permeate the mass of the plastic. Also, if water cannot easily enter by diffusion, osmotic rupture of the plastic will be minimized.[7] Utilization of the carbon content of the plastic will therefore be confined to the exterior surface. This more restricted attack by microorganisms will also be dependent on the physical properties of the surface of the plastic. Thus, a low contact angle will reduce the wettability by water as well as the rate of biodegradation. Similarly, the smoothness of the plastic surface will determine the true area accessible to

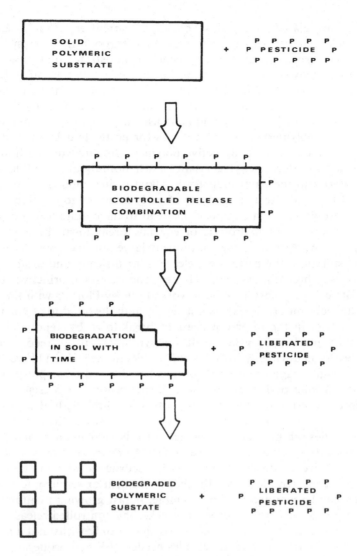

FIGURE 1.   Pictorial representation of the release of pesticide from
a biodegradable substrate.

attack and minute depressions present will act as receptacles for the water needed for microorganism growth.

For all of these reasons, the naturally occurring polymers are widely regarded as quite biocompatible, whereas synthetic polymers tend to be somewhat suspect. As a consequence, more research has been carried out on the former which are, in addition, available in low-cost abundance as part of the waste products of our society.

III. VARIANTS WITHIN THE GENERAL SYSTEM: COVALENT LINKAGES

The forerunner of controlled release combinations prepared by covalently linking a pesticide to a substrate was the inorganic ester Fenone, tris(2,4-dichlorophenox-ethyl)phosphite,[8] which was introduced in 1956. In this case there is a multistep mechanism controlling the availability of the actual herbicide, 2,4-dichlorophenoxyacetic acid (2,4-D), to the plant. First, the triester must hydrolyze and there are three different rate constants associated with this release. Thereafter, the 2-(2,4-dichloro-phenoxy)ethanol which is released must be oxidized through the corresponding alde-hyde to the active herbicide acid. This sequence demonstrates that modifications of the persistence and activity of a pesticide can be produced by reversible deactivation.

In spite of the theoretical advantages of this multistep release, the controlled release product was, in fact, a liquid. This physical characteristic led in actual applications to wasteful dissipation by facile volatilization and leaching. These shortcomings presaged the potential use of polymeric esters as vehicles for pesticide delivery and as early as 1960 Faerber described fungicidally active polymers prepared by the homo- and co-polymerization of *m*-chlorophenyl acrylate.[9] Such nonmigratory mildewicides have long been of interest to the paint industry.[10] Alkyd resins, modified by the telomeric inclusion of 2,4-D, were also the products of early attempts to develop sustained re-lease systems.[11] However, such compositions tend to have quite low moisure equili-rium contents. The rates of hydrolytic or enzymic breakdown therefore tend to be correspondingly slow. Although the creation of large surface areas by grinding can ameliorate this situation, the pattern of release is intrinsically undesirable. This is be-cause the alkyd is a three-dimensional polymer. The statistical principles of the mode of formation of such polymers have been worked out by Flory,[12] with the recognition that the instantaneous onset of gelation is a key point in the buildup of macromolecular structures. Similarly, in the reverse process of breakdown by depolymerization, the extent of reaction marked by gelation during synthesis will be recognized by cata-strophic disintegration during hydrolysis. This results in the sudden liberation of many low molecular weight fragments, which in turn should lead to a splurge of herbicide availability after a long period of minuscule dispensation. At present, the utility of such a behavioral pattern is not apparent because it is a shibboleth of controlled release philosophy that a constant rate of delivery to replace losses of the bioactive materials is the epitome of desirability. The correctness of this view is questionable and argu-ments for on-off systems can certainly be made. Moreover, in the plant kingdom the opinion that the delivery rate should increase in accordance with the size of the plant being treated has considerable merit. Indeed, in tropical or semitropical countries an increasing rate may be mandatory. For example, if fast-growing plantations of Meli-aceae are to be protected from shootborer attacks, the controlled release formulation has to initially protect a very small seedling which can only withstand a small dose of the somewhat phytotoxic systemic insecticides needed.[13] When properly protected this seedling can grow several feet in the course of a few months and the resultant larger tree then requires a greater amount of pesticide to prevent shootborer damage. Clearly, an increasing rate of delivery is called for under these circumstances.

In fact, the established long-term effectiveness of pesticides chemically attached to biodegradable substrates[14] may be because the amount liberated is enhanced by biolog-ical breakdown late in the biocidal life of the combination. That is to say, the degra-dation of the substrate may proceed at three different rates relative to the cleavage of the pesticide substrate bond. First, the depolymerization rate can be substantially less than that of the pesticide release rate. The active ingredient will then be supplied to the environment in ever-decreasing amounts approximating to first order reaction ki-netics. Second, the two rates may be roughly equivalent. This condition implies that the ratio of active ingredient to the substrate polymer remains relatively constant. Ac-cordingly, the release rate of the pesticide must continue unchanged. The third possi-bility is that the depolymerization of the substrate can proceed at a rate which is much faster than that of the scission of the substrate = pesticide bond. Under these condi-tions, the concentration of biocide moieties on the backbone will actually increase. Therefore, the rate of release should increase correspondingly.

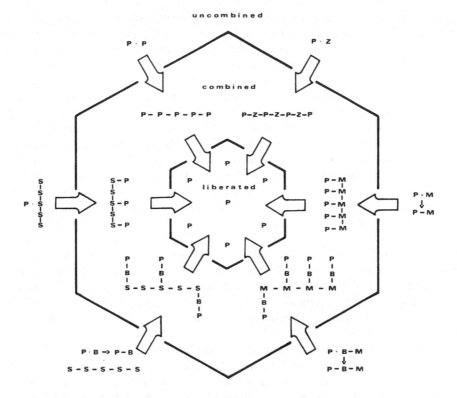

FIGURE 2. Summation of alternative strategies for the construction of chemically bonded controlled release systems based on pesticide-polymer permutations. The letters P, S, B, M, and Z denote pesticide, preexisting polymeric substrate, bridging entity, monomer and co-monomer, respectively.

The effectiveness of these bioactive, biodegradable polymer derivatives is governed by the rate of pesticide release, which in turn is controlled by: (1) the nature and stereochemistry of the pesticide-polymer bonds, (2) the presence and nature of other pendant groups, (3) the level of substitution and any other factors influencing the hydrophobicity and (4) the accessibility of pesticide-polymer bonds to the degrading agent. Furthermore, within limits, the degree of polymerization of the substrate material and its extent of cross-linking will also be important.

With these constraints in mind, some possible structures of chemically bonded controlled release biocide combinations are depicted in the Kekulean diagram of Figure 2. These entities can actually be subdivided into two main classes according to the type of biocide-substrate linkage: (1) combinations containing a bond directly from the active entity to the polymer and (2) combinations containing a functional intermediary between the active entity and the substrate.

IV. DIRECT BONDING OF PESTICIDES TO POLYMERS

A requirement for the preparation of such materials is that the polymer and pesticide each contain accessible and mutually reactive functional groups. Subsequently, the bond created in the reaction must degrade under the environmental conditions of usage to release the original active compound. The substrate-biocide bond must therefore break more readily than any bond in the biocide. The most successful controlled release combinations to date utilize ester, amide or anhydride linkages.[15]

Phenoxy carboxylic acids and the hydroxy-rich natural polymers, such as cellulose, chitin, chitosan, lignin, starch, alginic acid or lignocellulosic bark, make convenient reaction partners for the synthesis of practical controlled release systems useful as herbicides for both land and sea plants.[16] All of the resultant products summarized in Table 1 contain esterified herbicide moieties, but mixed anhydrides of phenoxy acids with cellulosic and lignin carboxylic acid groups may also be present. In use, these less stable linkages will give a fast burst of release which is desirable since the minimum effective level for biological response is thereby rapidly attained. The traditional organic synthesis route to these combinations has been via an acid chloride intermediate,[17] and Allan and co-workers have comprehensively described the combination of pesticides with natural polymers.[18] Recent University of Washington research has shown that similar combinations can be prepared without recourse to the use of reactive intermediates.[19, 20] In general, heating carboxylic acid herbicides above their melting points in the presence of natural polymers yields covalently condensed products. More specifically, melting 2,4-D at 150° together with Douglas-fir (*Pseudotsuga menziesii*) bark, and maintaining this temperature for 3 hr, yields a product which, when subjected to hydrolysis, affords 20% of its weight as herbicide. Exchange methods for the synthesis of these polymer esters have also been examined, and heating 2,4-D with cellulose acetate at 150° for 3 hr provides a controlled release combination containing 33% of chemically linked herbicide.[19] Alternatively, the anhydride form of carboxylic acid herbicides can be used to effectively esterify polymers,[2] although the preparation of such reagents is not particularly facile.

Mixed anhydrides of the pesticide and polymer can, however, be easily synthesized and are feasible controlled release systems (Table 2). For example, poly(acrylic acid) (Carbopol 941®, B. F. Goodrich Chemical Co., Cleveland, Ohio) provides ample opportunity for anhydride formation with the acidic herbicide, 2,4-D. In addition to this type of pendant anhydride, poly(sebacic anhydride) terminated with 2,4-D units was prepared as an example of an intrachain aliphatic polyanhydride. Concomitantly, poly[1,3-*bis* (*p*-carboxyphenoxy) propane] anhydride, also chain-terminated with 2,4-D, was studied to compare the effects of introducing aromaticity and hydrophobicity vicinal to the anhydride bonds. These controlled release herbicides were collectively studied under laboratory bioassay conditions.[21]

The results obtained showed that the duration of herbicidal effectiveness attained with the poly(acrylic acid) and poly(sebacic acid-) anhydride combinations was not significantly different from that achieved by simply applying free 2,4-D. However, the vicinal enclosure of the anhydride link by hydrophobic aromatic moieties did increase the duration of activity by 1.75 times in comparison to 2,4-D alone. Clearly then, the immediate environment around the labile bond has a profound effect upon the degradative release of the active herbicide. It is important to realize that these results do not mean that the aliphatic polyanhydrides could not make acceptable controlled release combinations. They may mean that the best application for labile anhydrides will be in situations where short-term control is all that is necessary. To put it another way, it is now known that the stringent conditions of a laboratory bioassay accelerate degradation of hydrolyzable bonds by 4 to 10 times relative to actual field conditions.[20] Therefore, in the drier regions of the U.S., such as the southeast and west, such anhydride combinations could offer a diversity of release rates.

Comparable examples of the modification of synthetic polymers (Table 3) are provided by the esterification of poly(vinyl alcohol) with the acid chloride derivatives of 2,4-D, 2,2-dichloropropionic acid (dalapon)[22] and 4-chloro-2-methylphenoxyacetic acid (MCPA),[23] to give products containing about 50% by weight of liberatable herbicide. Herbicidal amide derivatives of poly(ethyleneimine) containing 78% of releasable active ingredient (a.i.) were readily prepared by refluxing the polyamine with 2,4-D or the methyl ester of 2,4-D, in dry benzene.[22] Additional examples of bioactive

TABLE 1

Composition of Herbicide - Natural Polymer Combinations

Reactants	Controlled release combination	Herbicide content[a](%)
2,4-Dichlorophenoxyacetic acid (2,4-D) + α-cellulose		4.1, 4.9, 7.3, 15.3, 17.5, 24.2, 31.8
2,4-D + bark		5.3, 9.7, 12.3, 18.9, 19.7, 27.4, 32.4, 33.2, 34.7
2,4-D + kraft lignin		15.1, 21.8, 39.0
2,4-D + chitin		3.0, 7.9, 9.9, 14.8, 21.8, 23.4
2,4-D + crab shells		6.1, 9.2, 22.3
2,4-D + fish waste		7.0, 15.0, 19.5, 24.5, 28.0

TABLE 1 (continued)

Composition of Herbicide - Natural Polymer Combinations

Reactants	Controlled release combination	Herbicide content[a](%)
2,4-D + chitosan		21.0, 26.0, 27.0, 31.0, 32.0, 38.0, 42.0, 44.0, 45.0, 46.0, 53.0, 54.0
4-(2,4-Dichlorophenoxy) bu-tyric acid (2,4-DB) + bark		20.8, 30.9, 36.8, 50.0
2,4-DB + kraft lignin		29.5, 39.3
2,4,5-Trichlorophenoxy-acetic acid (2,4,5-T) + bark		17.5, 23.8
2,4,5-T + α-cellulose		28.6, 38.1

TABLE 1 (continued)

Composition of Herbicide - Natural Polymer Combinations

Reactants	Controlled release combination	Herbicide content[a](%)
2,4,5-T + kraft lignin	Cl-substituted phenyl ring, $O-CH_2C(=O)-O-$kraft lignin	37.5, 50.0
4-(2,4,5-Trichlorophenoxy) butyric acid (2,4,5-TB) + bark	Cl-substituted phenyl ring, $O-(CH_2)_3-C(=O)-O-$bark	11.5
2,4,5-TB + kraft lignin	Cl-substituted phenyl ring, $O-(CH_2)_3-C(=O)-O-$kraft lignin	21.9
4-Chloro-2-methylphen-oxyacetic acid (MCPA) + bark	CH_3, Cl-substituted phenyl ring, $O-CH_2-C(=O)-O-$bark	29.8, 51.0
MCPA + kraft lignin	CH_3, Cl-substituted phenyl ring, $O-CH_2-C(=O)-O-$kraft lignin	23.0, 38.5
4-(4-Chloro-2-methylphen-oxy) butyric acid (MCPB)	CH_3, Cl-substituted phenyl ring, $O-(CH_2)_3-C(=O)-O-$bark	8.2, 28.9, 43.9
MCPB + kraft lignin	CH_3, Cl-substituted phenyl ring, $O-(CH_2)_3-C(=O)-O-$kraft lignin	12.9, 41.8
Trichloroacetic acid (TCA) + bark	$Cl_3C-C(=O)-O-$bark	17.9

TABLE 1 (continued)

Composition of Herbicide - Natural Polymer Combinations

Reactants	Controlled release combination	Herbicide content[a](%)
TCA + kraft lignin	$\begin{array}{c}\text{Cl} \quad \text{O} \\ \mid \quad \parallel \\ \text{Cl–C–C–O–kraft lignin} \\ \mid \\ \text{Cl}\end{array}$	6.5, 9.7
TCA + sawdust	$\begin{array}{c}\text{Cl} \quad \text{O} \\ \mid \quad \parallel \\ \text{Cl–C–C–O–sawdust} \\ \mid \\ \text{Cl}\end{array}$	11.9
2,2-Dichloropropionic acid (dalapon) + bark	$\begin{array}{c}\text{Cl} \quad \text{O} \\ \mid \quad \parallel \\ \text{CH}_3\text{–C–C–O–bark} \\ \mid \\ \text{Cl}\end{array}$	20.4, 33.0
Dalapon + kraft lignin	$\begin{array}{c}\text{Cl} \quad \text{O} \\ \mid \quad \parallel \\ \text{CH}_3\text{–C–C–O–kraft lignin} \\ \mid \\ \text{Cl}\end{array}$	10.6
Dalapon + sawdust	$\begin{array}{c}\text{Cl} \quad \text{O} \\ \mid \quad \parallel \\ \text{CH}_3\text{–C–C–O–sawdust} \\ \mid \\ \text{Cl}\end{array}$	13.7
Dalapon + chitin		33.6
4-Amino-3,5,6-trichloro-picolinic acid (picloram) + bark		11.0

TABLE 1 (continued)

Composition of Herbicide - Natural Polymer Combinations

Reactants	Controlled release combination	Herbicide content[a](%)
Picloram + kraft lignin		7.6
Picloram + sawdust		8.6
2-(2,4,5-Trichlorophenoxy) propionic acid (silvex) + bark		8.9, 18.2, 30.9, 32.5
Silvex + kraft lignin		12.3, 28.2, 42.2
3,6-Dichloro-*o*-anisic acid (dicamba) + fish waste		23.0, 30.0, 31.0, 32.5

[a] Expressed as the weight of free bioactive agent liberatable by hydrolysis.

TABLE 2

Composition of Herbicide - Anhydride Polymer Combinations

Reactants	Controlled release combination
2,4-Dichlorophenoxyacetic acid (2,4-D) + 1,3-*bis* (*p* - carboxyphenoxy) propane	
2,4-D + sebacic acid	
2,4-D + poly(acrylic acid)	

TABLE 3

Composition of Ester- or Amide-Linked Herbicide - Synthetic Polymer Combinations

Reactants	Controlled release combination	Herbicide content (%)[a]
2,4-Dichlorophenoxyacetic acid (2,4-D) + poly(vinylalcohol)		42.8
4-Chloro-2-methylphenoxyacetic acid (MCPA) + poly(vinylalcohol)		56.1
2,2-Dichloropropionic acid (dalapon) + poly(vinylalcohol)		45.0
2,4-D + poly(ethyleneimine)		78.0
2,4-D + poly(ethyleneimine) via methyl ester of 2,4-D		76.0

[a] Expressed as the weight of free bioactive agent liberatable by hydrolysis.

TABLE 4

The Effect of Varying the Ratio of Reactants and the Duration and Temperature of Reaction Upon the Extent of Combination of 2,4-D with Chitosan

Duration of reaction (hr)	Initial weight ratio of 2,4-D to chitosan					
	1:1		2:1		4:1	
	165°[a]	195°[a]	165°[a]	195°[a]	165°[a]	195°[a]
1	45[b]	31[b]	42[b]	44[b]	—	53[b]
3	21	32	34	46	53	54
5	27	26	43	31	38	34

[a] Reaction temperature.
[b] Content of herbicide to chitosan in controlled release combination, % w/w.

synthetic polymers which utilize labile ester linkages have been prepared by Jakubka and Busch[24] by the metathetical reaction of the triethylammonium salts of 2,4-dichlorophenoxyacetic, 4-chloro-2-methylphenoxyacetic, 4-phenylazophenoxyacetic, 8-quinolinoxyacetic, 2-(2,4-dichlorophenoxy) propionic, 2-(4-chloro-2-methylphenoxy) propionic, 4-(2,4-dichlorophenoxy) butyric, and 4-(4-chloro-2-methylphenoxy) butyric acids with the chloromethylated derivative of a divinylbenzene-styrene copolymer. The corresponding amide-linked analogs were produced by using dicyclohexylcarbodiimide to acylate the aminomethylated derivative of the styrenic copolymer with the same phenoxy acids.

More accessible controlled release forms based on amide linkages are exemplified by the combination of phenoxy herbicide acids with natural polymers such as the aminopolysaccharide, chitosan,[16] or fish wastes, which represent a proteinaceous polymeric material containing reactive amino and carboxylic acid groups. The reaction of 3,6-dichloro-*o*-anisic acid (dicamba) with proteinaceous fish waste yielded a biodegradable controlled release herbicide combination (37% a.i. by wt) that suppressed the growth of vegetation in bioassay pots five times longer than did an equivalent amount of the free herbicide.[20] In addition, covalent attachment of 2,4-D to chitosan by simple heat-induced condensation afforded products containing as much as 45% by weight of releasable herbicide. The products obtained by employing various reaction times and temperatures are catalogued in Table 4. The practicality of chitosan as a substrate for controlled release systems will be augmented as its availability from crab and shrimp shells is extended by utilizing fungal hyphae and other sources.[25]

The relative positioning of the precursor groups of the amide function can, of course, be transposed if the herbicide contains an amino group. The acid moiety would then be pendant from the polymer. An interesting example of this type of controlled release entity was synthesized by heating, in the dry state, lignin sulfonic acid which had been neutralized with the herbicide, 3-amino-1,2,4-triazole (amitrole). The polymeric product, which may contain sulfonamide or imine linkages, was free of uncombined amitrole, and yet exhibited the characteristic herbicidal activity of this chemical in pot tests.

Amide derivatives of totally synthetic polymers can also be prepared, but the range of substrate polymers available is not large with polyethyleneimine (PEI), really the only commercially significant material. Poly(ethyleneimine-*S*-alkyldithiocarbamates) which act as controlled release herbicides and fungicides have also been described.[26,27] These polymers decompose to release a host of ill-defined sulfur containing substances which show biocidal effects.

V. IONIC- AND CHELATE-BONDED PESTICIDE-POLYMER COMBINATIONS

The bioactive agents considered in the foregoing discussion were attached to the polymer backbone by covalent, hydrolyzable bonds. There is, of course, no absolute requirement that this highest energy type of bond be the linkage used. Conceptually at least, it should be possible to develop controlled release delivery systems based on the other recognized types of lower energy chemical bonds. These comprise the very low energies ($\Delta H \simeq 1$ kcal/mol) associated with van der Waals forces, which under special circumstances become hydrogen bonds (ΔH, 1 to 10 kcal/mol). Beyond these on the bond energy spectrum, lie ionic bonds (ΔH, 10 to 15 kcal/mol) and coordinate or chelate linkages (ΔH, $\simeq 50$ kcal/mol). Actually, pure pesticides absorbed on carriers represent a very time-abbreviated form of controlled delivery. Certainly, for example, an insecticide absorbed onto corn cobs would be expected to provide a period of protection of longer duration than the same insecticide applied in acetone solution to the soil. Similarly, a strongly hydrogen-bonding pesticide could be quite tenaciously retained by a siliceous clay carrier were it not for the exceptional competition provided by the hydrogen bonding capabilities of water. On the other hand, bioactive agents linked ionically to a polymer should be able to resist the releasing action of water more effectively. Presumably, the rate of delivery of the pesticide would be determined by the dissociation of the polymer salt. This precept was investigated by Friedhoff[21] using the ionic salts of 2,4-D in an effort to secure short-term controlled release herbicidal action. Poly(ethyleneimine), chitosan, and a tertiary aminomethylated cross-linked poly-(styrene) resin (Dowex® 1-X8) were used as the polybasic materials for salt formation with 2,4-D. The results of these studies indicate that only with PEI fully substituted with 2,4-D was any increase in herbicidal effectiveness observed in bioassay evaluations. This increase amounted to 25%. These results suggested that the most important parameter for these combinations was the decrease produced in the water solubility of the herbicide. This view was borne out by examination of the 2,4-D salt of Dowex® 1-X8. The duration of herbicidal effectiveness was doubled in this case. This behavior is in contrast to that of the analogous salt of the herbicide and the more water soluble polymer, chitosan. In this use, the duration of herbicidal effectiveness was increased by only 30%. Clearly, the degree of increased herbicidal persistence attainable with this type of controlled release formulation is substantially dependent on the water sensitivity of the polymeric salt.

However, decreasing the solubility of 2,4-D by formation of its calcium salt substantially reduced its herbicidal activity to below that of the free acid. Apparently, the solubility product for the calcium salt of 2,4-D is below that necessary to permit the buildup of 2,4-D concentrations to a level sufficient to inhibit lettuce seed germination. Moreover, any calcium ion concentration in the soil will serve to further inhibit the liberation of 2,4-D by exerting a depressant common ion effect. Nonetheless, other metal salts of 2,4-D may possess the appropriate hydrolysis constants to provide a useful herbicidal level of release. Thus, the uncharacterized controlled release "polymers" claimed by Beasley and Collins[28, 29] are, most likely, simply iron salts which may be fortuitously herbicidal.

Beyond simple salt bonds on the energy scale lie coordinate or chelate bonds. With this type of bonding, it is possible to create highly stable, complex, organometallic structures and there is a vast body of background chemical information defining the stability constants of such entities.[30] Strangely enough, controlled release delivery systems based on the cleavage of coordinate bonds have been largely ignored. An exception to this neglect[31] is provided by the polymeric iron and aluminum chelates (I) of the herbicide 4-amino-3,5,6-trichloropicolinic acid (picloram).

1

These materials were formed by first saturating a commercial chelate resin (Dowex®
A-1, Dow Chemical Co., Midland, MI) with either Fe^{3+} or Al^{3+} ions. Subsequent elution of a column of the resin with a solution of the herbicide led to its capture by the
metallic sites. This particular herbicide has a tendency to become ineffective because
it can easily be leached down into the subsoil. Therefore, the iron and aluminum chelates were compared with the parent material in a soil column percolation test.[32] Examination of 1 in. segments of the soil column after elution with water showed that,
while the free picloram, as the potassium salt, moved very rapidly down in almost a
plug flow, both chelates gave a totally different distribution of herbicide. The bioactive
Fe^{3+} complex was largely confined to the upper regions of the column. In contrast, the
Al^{3+} coordinate system was so stable that no bioactivity could be detected at any level
when the soil samples were seeded with cucumbers. These results demonstrate that
chelated controlled release systems are feasible, although further research is needed to
establish the utility of more readily available and lower cost backbone polymers which
contain suitably oriented potential chelating sites.

VI. POLYMERIZATION OF PESTICIDE-MONOMER COMBINATIONS

The questions of selection, availability, cost, and ease of reaction of backbone polymers have repeatedly stimulated the idea of directly synthesizing polymeric pesticides
from monomeric derivatives. This approach was initially investigated by Faerber[9] and
by Kostanyan and Dovlatyan.[33] The latter researchers prepared poly(vinyl-2,4-dichlo-
rophenoxyacetate) from the corresponding vinyl monomer. Hydrolytic release of the
phenoxy acid from this combination is effectively nonexistent because of the hydrophobicity of the polymer. This is not too surprising because the rate of release of the
active ingredient is related to the rate of scission of the covalent bond between the
grafted pesticide and the polymer substrate. This, in turn, is regulated by the strength
and chemical nature of that bond, as well as the stereochemical and chemical properties
of the vicinal polymer backbone. Of course, some cleavage of the pesticide-polymer
link may occur within the interior of the combination, and diffusion thereby may also
play a role in the ultimate liberation of the bioactive component. However, only covalent bond breakage as exemplified by the hydrolysis of the polymeric esters of pesticides will be considered in this discussion.

Since a hydrophobic polymer can essentially deactivate an ester of a grafted pesticide
acid towards hydrolysis, increasing the hydrophilicity of the backbone ought to just
as easily enhance hydrolytic scission, and this has been demonstrated by both Neogi[23]

and Wilkins.[22] The controlled release combinations of MCPA listed in Table 5 illustrate this point quite well.[23] Thus, poly(vinyl-4-chloro-2-methylphenoxyacetate) showed no biological activity when bioassayed. However, the introduction of hydrophilicity by copolymerization of the vinyl ester of the pesticide with 21% by weight of acrylic acid imparted a herbicidal character to the combination. Similarly, the homopolymerization of the hydrophilic glycidyl ester of MCPA produced a bioactive polymer. Neogi also homopolymerized 2,3,5-trichloropyrid-4-yl methacrylate and found that the polymer possessed no biological activity. However, a bioassay showed positive activity in the product obtained by the copolymerization of the heterocyclic monomer with 42% by weight of acrylic acid.

Similar tests and conclusions were reported by Wilkins.[22] These are also catalogued in Table 5. The homopolymers of the vinyl esters of 2,4-D, 2,4,5-trichlorophenoxyacetic acid (2,4,5-T), and 4-(2,4,5-trichlorophenoxy)butyric acid (2,4,5-TB) showed no capacity to inhibit the germination of lettuce seeds in a soil bioassay. However, the copolymer of vinyl (2,4-dichlorophenoxyacetate) with 25% by weight acrylic acid was capable of suppressing germination.

The detail of the factors controlling release from these types of polymers have since been extensively investigated at Wright State University[34] and Chapter 3 in this book is devoted to the topic.

VII. CONDENSATION POLYMERIZATION OF BIFUNCTIONAL PESTICIDES

In all of the pesticide-polymer combinations thus far discussed, there is a backbone of polymer which is not biologically active. From one point of view, this can be regarded as wasteful, and a recent patent by Allan and Neogi[35] has described how pesticides containing more than one functional group can be converted into polymers which are entirely composed of bioactive material. This system would liberate the active ingredient through the degradation of the main polymer chain, rather than through pendant group scission (Figure 2). Since release of the active biocide can only occur through endwise degradation, it can be predicted that increasing the degree of polymerization (D.P.) of linear polymers should result in a diminished release rate. Thus, the rate of release depends upon the number of end groups, which is inversely related to the D.P. of a straight-chain polymer. This implies that branched structures, which have more chain ends than their linear counterparts, should exhibit a more rapid liberation of their active component. Regardless of the particular complexities of the backbone, the linkages must be hydrolyzable, and from a study of the various types listed in Table 6, Neogi concluded that aromatic polyamides were generally too stable to be useful.[23] An exception to this generalization is poly(3,5,6-trichloropicolinamide), which is a highly effective herbicide. Presumably, the heterocyclic nitrogen atom in the aromatic ring provides some anchimeric assistance to the hydrolyzing species. Additional examples of these types of bioactive polymers have been described by Akagane and Allan in a series of Japanese publications[36,37] dealing with antifouling compositions. For example, dehydrative fusion of monosodium arsenate,[37] first at 250° for 1 hr and then at 700° for 6 hr, alone or in the presence of elemental sulfur, afforded the inorganic polyanhydrides II and III.

$$
H \left[\begin{array}{c} O \quad\ O \\ \| \quad\ \| \\ -O-As-O-As- \\ | \qquad | \\ O^-Na^+ \ \ O^-Na^+ \end{array} \right]_n OH
\qquad\qquad
H \left[\begin{array}{c} O \quad\ O \\ \| \quad\ \| \\ -O-As-S-As- \\ | \qquad | \\ O^-Na^+ \ \ O^-Na^+ \end{array} \right]_n OH
$$

II   III

TABLE 5

Composition of Homo- and Co-Polymers of Derivatized Biocides

Vinyl (2,4-dichloro-phenoxy)acetate

Vinyl (4-chloro-2-methyl-phenoxy)acetate

Vinyl (2,4,5-trichloro-phenoxy)acetate

Vinyl 4-(2,4,5-trichloro-phenoxy)butyrate

Glycidyl (4-chloro-2-methyl-phenoxy)acetate

TABLE 5 (continued)

Composition of Homo- and Co-Polymers of Derivatized Biocides

Reactants	Controlled release product

N-(4-Chloro-2-methyl-
phenoxy)acetylaziridine

Pentachlorophenyl-
methacrylate

2,3,5-Trichloropyrid-
4-yl methacrylate

N-Hydroxymethyl(2,4-di-
chloro-phenoxy)acetamide

TABLE 5 (continued)

Composition of Homo- and Co-Polymers of Derivatized Biocides

Reactants	Controlled release product
N-Hydroxymethyl-N′-(di-chlorophenoxy)urea	
Vinyl (2,4-dichlorophen-oxy)acetate + acrylic acid	
Vinyl (4-chloro-2-methyl-phenoxy) acetate + acrylic acid	

TABLE 5 (continued)

Composition of Homo- and Co-Polymers of Derivatized Biocides

Reactants	Controlled release product
2,3,5-Trichloropyrid-4-yl methacrylate + acrylic acid	
7-oxabicyclo[2,2,1]heptane-2,3-dicarboxylic acid + poly (oxypropylene-diamine)	

The time dependency of the hydrolytic degradation of the sulfur-free polymer, from an initial weight W_o, experimentally follows (Figure 3) the theoretical relationship

$$t = (1/k) \ln(1/1 - L^{1/2})$$
(1)

where k is the reaction rate constant and

$$L = (W_o - W)/W$$

is the relative weight loss at an elapsed time, t, when the polymer weight has decreased to W. Although the antifouling properties of the arsenic-acid homopolymers were initially excellent, they rapidly deteriorated with time, whereas the copolymers with a sulfur monomer content of 5 to 10% remained effective for the 30-day duration of the test. Comparable excellent antifouling performance could be obtained both from the homopolymer, poly(phenarsazine) (DP $\simeq$ 3), and from the copolymer (DP $\simeq$ 8) of 1-hydroxy-2-(4-hydroxyphenyl)-3-oxido-4,5-dimethylimidazole and terephthalic acid. In either case, the degree of polymerization had to be kept low so that a sufficiency of polymer chain ends was available to provide the necessary hydrolyzable monomer moieties for satisfactory performance.

VIII. INDIRECT BONDING IN CONTROLLED RELEASE COMBINATIONS

In spite of the attractive compactness of these controlled release condensation polymers, there are often situations where a relatively low content of active ingredient is mandatory to facilitate distribution. It is not by accident that most granular agricultural formulations contain only 10 to 20% pesticide. This consideration makes selection of low level attachment to low-cost substrates difficult to resist. Frequently, nonetheless, the most desirable selections of biocide and substrate contain functional groups which do not lend themselves to direct covalent attachment. This situation arises from

TABLE 6

Composition of Herbicide Homopolymers

Reactant	Controlled release product
4-Amino-3,5,6-trichloro-picolinic acid (picloram)	
7-Oxabicyclo[2,2,1]heptane-2,3-dicarboxylic acid (endothall)	
Tetrachloroterephthalic acid	
3-Amino-2,5-dichlorobenzoic acid (chloramben)	

either the inability of the reactants to form a bond whose intrinsic strength will permit a useful release rate, or because the substrate and the pesticide are not mutually reactive. This obstacle can often be overcome by interposing a multifunctional entity to act as a bridge between the incompatible groups (Figure 2).

The bond between the pesticide and bridging compound must be easily broken to provide a suitable release rate of the active ingredient. Generally, this bond is less stable to environmental influences than the link between the bridging compound and the substrate. If the bridge and biocide are released in twain, there is the danger of the inactivated, but mobile, pesticide being lost from the target area. However, release of an inactive form of the pesticide that can be carried to the pest before degradation regenerates the active agent could be an advantage. Once again, the underlying principles can be observed in nature. As previously mentioned, some plants dispense modified allelopathic compounds as mobile glycosides which hydrolyze to release the active material after entering the root zone of the competing plants.[4]

The preparation of synthetic, bridged formulations usually involves the initial reaction of equimolecular amounts of pesticide and bridging compound. Subsequently, this product is reacted with the substrate by means of the remaining active moiety. Polyfunctional compounds such as cyanuric chloride, phosphoryl chloride, and silicon or titanium tetrachloride are attractive in this respect, since more than one equivalent of pesticide can be linked to the polymer using only one equivalent of the bridging

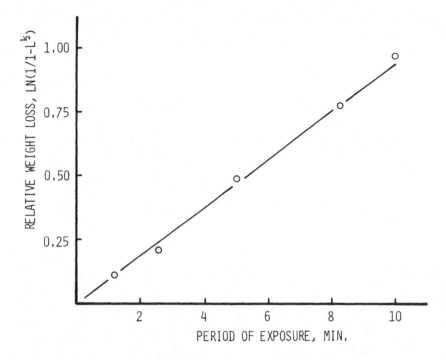

FIGURE 3.   The time dependency of the aqueous dissolution of the polyanhydride of monosodium arsenate.

entity.[38] The attachment reactions can, of course, be carried out in the reverse order.[39] Both of these techniques have been used by Allan and his various associates to interpose a variety of bridging molecules and prepare many biodegradable controlled release systems (Table 7). Naturally, care must be exercised to avoid side reactions which do not attach biocide and which may only cross link the polymer substrate.[40]

An example of the interposition of a polyfunctional bridge is provided in a study by Allan and Halabisky[39] where the fungicide pentachlorophenol was covalently bonded to a softwood kraft pulp which had previously been reacted with the trifunctional cyanuric chloride. The goal in this research was to determine the efficacy of this controlled release system as a prophylaxis for cellulosic materials against biological attack. Protection from such attack can usually be provided by simple impregnation of the cellulosic substrate with one or more of a host of known fungicides.[41-43] Unfortunately, this procedure is inadequate in those situations where the impregnated materials are subject to severe water leaching. The consequent loss of a sufficient quantity of protectant permits the cellulose to become subject to microbial attack. Clearly, once the pentachlorophenol has been bonded to the substrate via the cyanuric chloride bridge, lixiviation can only occur after hydrolysis liberates the fungicide, and long-term protection against the damage caused by cellulose feeding organisms can therefore be envisaged. This, in fact, transpired to be the case since untreated pulp sheets which had been placed in contact with the fungus *Lenzites trabea* for 7 weeks retained only 18% of their original strength, while the corresponding retention was 93% for the sheets containing the controlled release fungicide. Protection from the ravages of fungal deterioration can also be achieved by physical rather than chemical entrapment of a fungicide within the cellulose phase. This technique is based on the so-called "Jack-in-the-box" effect[44] whereby a polyelectrolyte in solution at its isoelectric point is first equilibrated with a microporous substrate. Cellulose fibers are noteworthy for their large internal surface area.[45] Subsequent pH adjustment of the solution within

TABLE 7

Composition of Bridged-Pesticide/Natural-Polymer Combinations

Bridging compound	Pesticide	Biodegradable substrate	Active ingredient (%)
Polybasic inorganic acid halides			
Phosphoryl chloride	2,3,5-Trichloropyridin-4-ol	Kraft lignin	10.5
POCl₃		Bark	10.5
	O,O-Dimethyl-S-(N-methylcarbamoyl) methyl phosphorodithioate	Kraft lignin	11.6
		Bark	6.5
	2,3,5-Trichloro-4-hydroxyethoxy pyridine	Kraft lignin	44.3
		Bark	28.8
		Ammonium lignin sulfonate	44.8
	4-Amino-2,3,5-trichloropicolinic acid	Kraft lignin	52.5
		Bark	45.5
Phosgene	2,3,5-Trichloro-4-hydroxyethoxy pyridine	Kraft lignin	12.4
Cl—°C—Cl			
Silicon tetrachloride	2,3,5-Trichloropyridin-4-ol	Kraft lignin	16.4
SiCL₄		Bark	58.3
Titanium tetrachloride	2,3,5-Trichloropyridin-4-ol	Kraft lignin	13.3
TiCL₄	2,3,5-Trichloro-4-hydroxyethoxy pyridine	Kraft lignin	26.3
		Bark	57.3
Thiophosphoryl chloride	2,3,5-Trichloropyridin-4-ol	Kraft lignin	41.0
$$\begin{array}{c}Cl\\|\\Cl-P=S\\|\\Cl\end{array}$$			
Phosphorus trichloride PC₃	2,3,5-Trichloropyridin-4-ol	Kraft lignin	38.0

Anhydrides

Pyromellitic dianhydride

2,3,5-Trichloropyridin-4-ol	Kraft lignin	7.1
	Bark	10.0

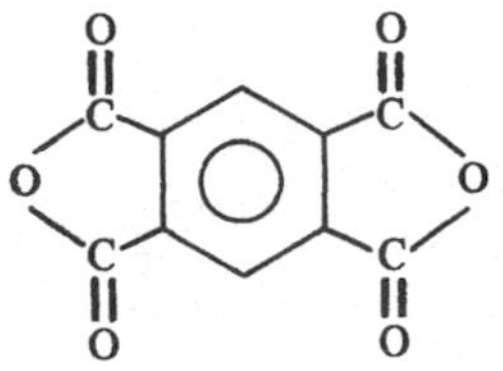

O,O-Dimethyl-S-(N-methylcarbamoyl) methyl phosphorodithioate

	Kraft lignin	2.9
	Bark	8.3

2,3,5-Trichloro-4-hydroxyethoxy pyridine

	Ammonium lignin sulfonate	29.5
	Bark	34.7

Styrene-maleic anhydride copolymer

2,3,5-Trichloropyridin-4-ol	Kraft lignin	10.0

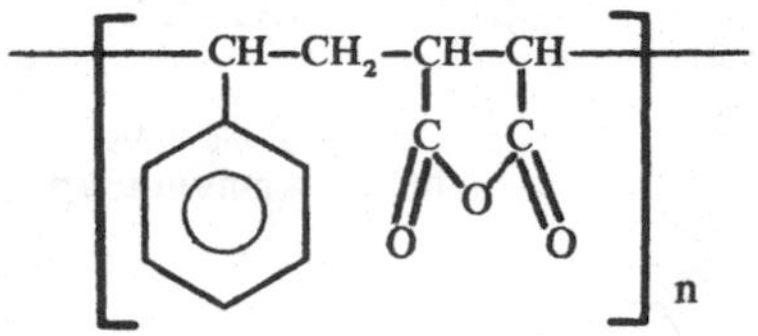

Aldehydes

Formaldehyde

4-Amino-2,3,5-trichloropicolinic acid	Bark	10.1

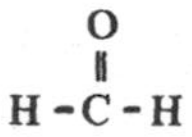

TABLE 7 (continued)

Composition of Bridged-Pesticide/Natural-Polymer Combinations

Bridging compound	Pesticide	Biodegradable substrate	Active ingredient (%)
Epoxides			
2,2-bis[p-(Glycid-3-oxy)phenyl]propane	4-Amino-3,5,6-trichloropicolinic acid	Bark	9.3
Unsaturated compounds			
Divinylsulfone	2,3,5-Trichloropyridin-4-ol	Kraft lignin	24.5
		Bark	32.0
Polybasic carboxylic acid chlorides			
Oxalyl chloride	2,3,5-Trichloro-4-hydroxyethoxy pyridine	Kraft lignin	19.4
		Bark	49.0

Polyisocyanate
Toluene diisocyanates

O,O-Dimethyl-S-(N-methylcarbamoyl) methyl phosphorodithioate	Kraft lignin	4.6
	Bark	7.8
2,3,5-Trichloro-4-hydroxyethoxy pyridine	Kraft lignin	21.6
	Bark	37.6
2,3,5-Trichloropyridin-4-ol	Kraft lignin	5.2
	Sawdust	9.5
	Bark	9.0

the fibers causes an exceedingly rapid expansion in the hydrodynamic volume, which traps the polyelectrolyte within the enclosing pores. Obviously, if the polyelectrolyte carries pendant fungicide moieties, these too will be trapped within the fiber. Akagane and Allan have synthesized several examples of this type of polymer by the covalent attachment of 2,4-D, pentachlorophenol or 4-amino-3,5,6-trichloropicolinic acid to either proteins or poly(ethyleneimine).[46] The attachment reactions used are summarized in Equations 2, 3, and 4. The use of acrolein as a low-cost, difunctional bridging agent is particularly worthy of note. After application of the fungicide-modified polyelectrolytes by the ''Jack-in-the-box'' technique, the cellulose fibers were found to be highly resistant to fungal breakdown.

$$R-\overset{\overset{O}{\|}}{C}-OH \ + \ CH_2=CH-CH=O \ \longrightarrow \ R-\overset{\overset{O}{\|}}{C}-O-CH_2-CH_2-CH=$$

$$O \ \xrightarrow{\ H_2N-R'\ } \ R-\overset{\overset{O}{\|}}{C}-O-CH_2-CH_2-CH=N-R' \tag{2}$$

$$(3)$$

$$R-\overset{\overset{O}{\|}}{C}-OH \ \xrightarrow{SOCl_2} \ R-\overset{\overset{O}{\|}}{C}-Cl \ \xrightarrow{H_2N-R'} \ R-\overset{\overset{O}{\|}}{C}-NH-R' \tag{4}$$

where

Recently, the idea of using a bulk commercial chemical as the bridge between pesticide and polymer has been receiving increased attention. In this vein, Nagayoshi combined the pesticides MCC (methyl-3,4-dichlorophenylcarbamate), methomyl [*S* - methyl-*N* -(methylcarbamoyloxy)thioacetamidate] and dimethoate (*O,O* -dimethyl-*S* - methylcarbamoylmethyl phosphorodithioate) with cellulose and starch by using toluene diisocyanate as the molecular linkage.[47] McCormick, Savage, and Hutchinson have also recognized the convenience of this commercially available bridge and have used it to attach the herbicide metribuzin, 4-amino-6-(1,1-dimethylethyl)-3-(methyl-thio)-1,2,4-triazin-5(4H)-one, to poly(vinyl alcohol).[48] This relatively costly backbone polymer is a tempting substrate because it is soluble and readily available in several molecular weight ranges. These attractions have led Schacht et al.[49] to employ it in an acid-catalyzed condensation with a precommercial herbicide, 2,6-dichlorobenzalde-hyde (DCBA). The product was a pendant cyclic polyacetal which was only subject to hydrolytic degradation at low pH values. In addition, an intrachain noncyclic polyace-tal of DCBA was prepared by the copolymerization of the biocide in the presence of an excess quantity of ethylene glycol. The resulting low molecular weight hydroxy-terminated copolymer was further condensed with 1,6-diisocyanatohexane to increase the size of this herbicidal macromolecule. These investigators then compared the rates of hydrolytic release afforded by the two herbicide-polymer controlled release combi-nations in aqueous dioxane and found that the backbone acetal copolymer of the DCBA liberated the active ingredient at a conspicuously greater rate than did the side-chain acetal of the poly(vinyl alcohol). Presumably, the relatively slower release of herbicide from the latter combination can be attributed to the bond stability and re-sulting deterred degradation induced by the pendant cyclization. Unfortunately, the bioactivity of neither type of herbicidal polyacetal has yet been reported.

Bridging technology is also of particular value when the entity to be attached is unstable or costly. This circumstance is exemplified by plant hormones, which Hilton[50] believes offer considerable potential in controlled release forms for the development of new agricultural techniques. In line with this prediction, the use of bridging com-pounds to prepare controlled release formulations of the plant hormones belonging to the cytokinin group has recently been reported.[51] These chemically bonded formula-tions are somewhat unusual in that the substrate to bridging compound bond was formed before connection to the pesticide. In addition, this second bond is the linkage which is hydrolyzed to allow release (Equations 5 and 6). Thus, phosgene acts as the molecular link to first produce a chlorocarbonate (IV) of the original polyhydroxy substrate. Subsequent condensation with the secondary amine of the cytokinin yields a decomposable urethane bridge (V). This linkage is hydrolytically destroyed in the plant environment and spontaneously eliminates carbon dioxide, thus regenerating the active cytokinin (VI).

(5)

(6)

where

IX. MULTIPLE PESTICIDE-POLYMER COMBINATIONS

Implicit in all of the foregoing is the idea that one type of biocide is attached to one type of polymer. This clearly need not be the case. For weed control in particular, there is frequently the need to apply more than one chemical, and synergistic effects are often recorded. Certainly, there is nothing to prevent the co-application of two distinct controlled release pesticide-polymer combinations. Nonetheless, situations where the materials could adventitiously separate can readily be foreseen. This might lead to erratic vegetation control. Moreover, combinations of pesticides are often more toxic than either biocide alone, and delivery systems which continuously provide both from an intimate admixture are likely to be more effective than conventional mixtures where one component may be dissipated and exhausted before its partner. Problems of incompatibility between pesticide-polymer pairs due to the physical differences, such as density, will also tend to be avoided if both pesticides are within the same controlled release form.

These views have been followed up in the area of antifouling paints by Akagane and Matsuura,[52] building on the former's research at the University of Washington. Typical examples of the type of polymer synthesized are the reaction products of commercial grade styrene-maleic anhydride copolymers with pentachlorophenol, phenylazo-β-naphthol, phenarsazine chloride, and tributyltin hydroxide. More recently, products with improved antifouling characteristics were obtained by the reaction of poly(acryloyl chloride) with two or more toxicants from the group consisting of phenarsazine chloride, tri- or penta-chlorophenol, tributyltin hydroxide, the novel 1-hydroxy-imidazoles[53] and their 3-oxido counterparts.[54] The mutual interactions of these toxicants on a single polymer gave better fouling control in a marine environment than mixtures of the corresponding polymers containing a single toxicant.[55]

A more complex herbicide system of this type based on low-cost lignin is described in a Japanese patent[56] (Equations 7 to 10) granted to Akagane. The product was synthesized by first etherifying the phenolic hydroxyl groups in kraft lignin with cyanuric chloride to obtain the corresponding dichloro-*s* triazinyl derivative (VII).[57] The active halogens were then replaced by the free amino groups present in the reaction product of 2,4-D and diethylenetriamine, which is formulated as VIII.

Finally, the as yet unreacted amino groups in the lignin complex (IX) were reacted with 3,4,5,6- tetrachlorophthalic anhydride and tributyltin hydroxide to yield the macromolecular herbicide which can be ideally represented as X. Although it will be a long time before such convoluted structures are used commercially, the synthetic flexibility available to the designer of controlled release systems is clearly demonstrated by their synthesis.

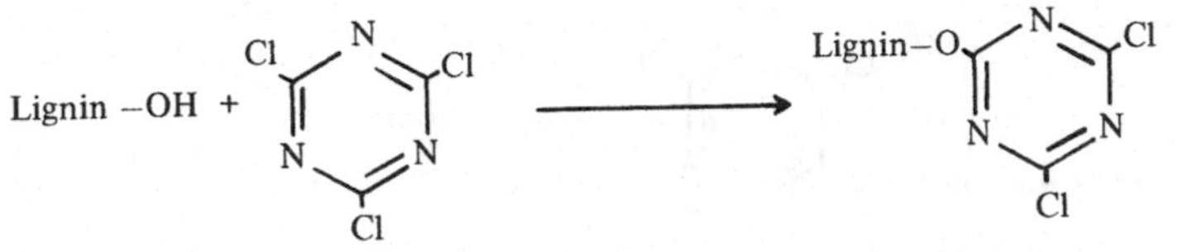

VII

(7)

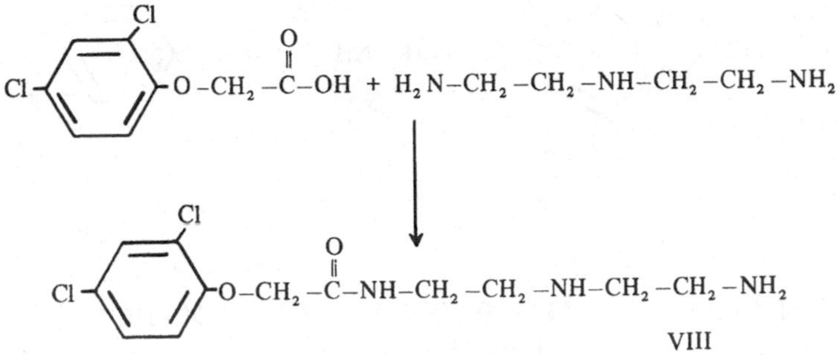

VIII

(8)

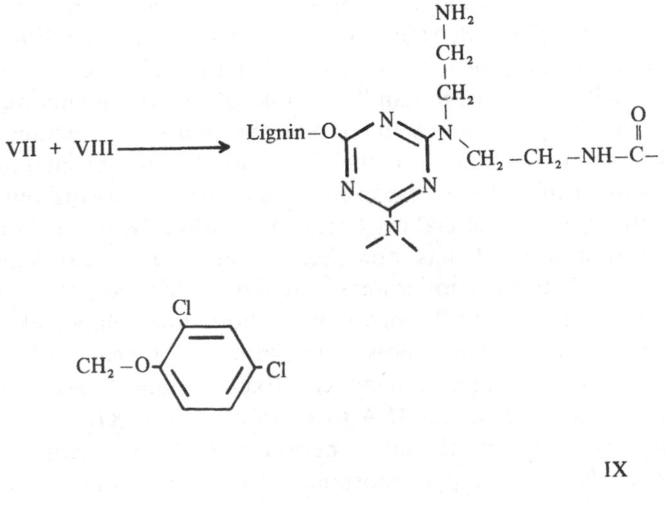

IX

(9)

(10)

X. THE *ab initio* SYNTHESIS OF PESTICIDE-POLYMER COMBINATIONS

It is also possible to synthesize controlled release biocide-polymer combinations without ever having the biological material in hand. This apparently paradoxical situation can exist when the pesticide is built up piece by piece starting from the polymeric substrate. This technique can be advantageous when the pesticide is difficult to synthesize, is unstable, or is hazardous to handle because of its extreme mammalian toxicity.

An example of this is provided by the organophosphate (XI), octamethylpyrophosphoramide (OMPA). This liquid systemic insecticide has a high mammalian toxicity and at one time was a candidate for use in conifers to nonlethally discourage the browsing of deer and elk. The commercial insecticide was, however, much too expensive and transient for this application. It has, nonetheless, been used as a rodenticide, where it functions by virtue of its anticholinesterase activity. However, it is not the OMPA itself that possesses this activity,[58] since over an hour must elapse after ingestion by the rodent before the characteristic poisoning symptoms appear. Other anticholinesterase-active compounds function immediately. Explanations offered for this slow action suggest that oxidation of the OMPA to an amine oxide (XIa) or a hydroxymethyl derivative (XIIb) occurs during the latent period, and the resultant electronic charge facilitates the hydrolysis of the pyrophosphate bond and ultimate reaction with chymotrypsin.

XIIa   XIIb

This hypothesis is supported by a considerable body of evidence, including reduction of the duration of the latent period by preoxidation to increase the rate of hydrolysis,[59] and the isolation of an oxygenated OMPA, which is easily hydrolyzed and which is still active after hydrolysis.[60] Evidence has also been presented that the conversion of OMPA to a cholinesterase inhibitor in the locust requires oxygen. It may be necessary, of course, that the cleavage of the pyrophosphate bond occur in vivo, but many simple phosphates are physiologically active, and no specific reference could be found to support this view. Moreover, it has been suggested that the usefulness of phosphoric acid esters in insecticides depends on their stability to hydrolysis.[61]

In any event, OMPA itself, and presumably its metabolites as acid anhydrides, are susceptible to hydrolysis. Based on the assumption that the hydrolysis products (XIII and/or XIVab) are physiologically active, it can be concluded that it should be possible to synthesize a polymer carrying pendant groups which would slowly hydrolyze to release the active pesticide component (Equation 11).

$$
\begin{array}{ccc}
\underset{Me_2N}{\overset{Me_2N}{\diagdown}}\hspace{-6pt}\underset{\underset{O}{\parallel}}{P}\hspace{-4pt}-\hspace{-4pt}O\hspace{-4pt}-\hspace{-4pt}\underset{\underset{O}{\parallel}}{P}\hspace{-6pt}\underset{NMe_2}{\overset{NMe_2}{\diagup}} & \xrightarrow{\text{HOH}} & \underset{Me_2N}{\overset{Me_2N}{\diagdown}}\hspace{-6pt}\underset{\underset{O}{\parallel}}{P}\hspace{-4pt}-\hspace{-4pt}OH
\end{array}
$$

XI   XIII

$$\downarrow O \hspace{5cm} \downarrow O$$

$$
\begin{array}{ccc}
\underset{Me_2N}{\overset{Me_2N}{\diagdown}}\hspace{-6pt}\underset{\underset{O}{\parallel}}{P}\hspace{-4pt}-\hspace{-4pt}O\hspace{-4pt}-\hspace{-4pt}\underset{\underset{O}{\parallel}}{P}\hspace{-6pt}\underset{N(Me)CH_2OH}{\overset{NMe_2}{\diagup}} & \xrightarrow{\text{HOH}} & XIII + HO\hspace{-4pt}-\hspace{-4pt}\underset{\underset{O}{\parallel}}{P}\hspace{-6pt}\underset{N(Me)CH_2OH}{\overset{NMe_2}{\diagup}}
\end{array}
$$

XIIb (or XIIa)   XIVb (or XIVa)

$$(11)$$

Kraft lignin was an obvious choice for the backbone polymer since it is low cost, readily available, and contains suitable functional groups for attachment of the phosphoramide moiety. From previous lignin-phosphorus research,[62] the following synthesis seemed reasonable (Equation 12).

$$
\text{lignin} \xrightarrow{\text{POCl}_3} \text{lignin} - O - \underset{\underset{O}{\parallel}}{P}\hspace{-6pt}\overset{\diagup Cl}{\underset{\diagdown Cl}{}} \xrightarrow{\text{Me}_2\text{NH}} \text{lignin} - O - \underset{\underset{O}{\parallel}}{P}\hspace{-6pt}\overset{\diagup NMe_2}{\underset{\diagdown NMe_2}{}}
$$

XV

$$(12)$$

This product (XV) should be capable of hydrolyzing to release a potential equivalent of OMPA. A similar synthesis leading to a closer relative of OMPA (XVI) could also be visualized using pyrophosphoryl chloride as depicted in Equation 13.[63,64]

$$
\text{lignin} \xrightarrow{\text{P}_2\text{O}_3\text{Cl}_4} \text{lignin} - O - \underset{\underset{O}{\parallel}}{\overset{\overset{Cl}{|}}{P}} - O - \underset{\underset{O}{\parallel}}{P}\hspace{-6pt}\overset{\diagup Cl}{\underset{\diagdown Cl}{}} \xrightarrow{\text{Me}_2\text{NH}} \text{lignin} -
$$

$$
O - \underset{\underset{O}{\parallel}}{\overset{\overset{NMe_2}{|}}{P}} - O - \underset{\underset{O}{\parallel}}{P}\hspace{-6pt}\overset{\diagup NMe_2}{\underset{\diagdown NMe_2}{}}
$$

XVI   (13)

Since the pyrophosphoryl chloride can be conveniently synthesized *in situ* by the reaction of phosphoryl chloride with the few percent adventitious moisture[65] invariably present in lignin preparations, the latter scheme is more attractive as an ultimate practical process and was the one chosen for study.

Before embarking upon the synthesis of the OMPA analog, it was demonstrated that prehydrolyzed OMPA applied to the soil could be picked up by young Douglas fir trees and translocated to the new needles. An extract of the tree was found to contain OMPA when examined by vapor phase chromatography. This surprising result indicated that either the tree or the vapor phase chromatographic column was reforming the pyrophosphoric acid derivative from the prehydrolyzed OMPA. Direct chromatography of the prehydrolyzed OMPA did, in fact, reform OMPA in the instrument, but this does not rigorously exclude similar reformation in the tree, and this point remains to be clarified. A key step in the carbon cycle in plants is the transformation of phosphates to pyrophosphates.

A laboratory investigation of the reaction of phosphoryl chloride with water and kraft lignin proceeded smoothly and, after the ultimate capping with dimethylamine, yielded a product which also can be represented as XVI. It should be appreciated that this is an idealized representation, and that other lignin-phosphorus-nitrogen combinations, such as XVII, are probably present.

$$
\begin{array}{ccc}
Me_2N & & NMe_2 \\
| & & | \\
lignin-O-P-O-P-O-lignin \\
\| & & \| \\
O & & O
\end{array}
$$

XVII

The merit of these views is illustrated by the analysis of the lignin-OMPA analog where all the added phosphorus, but only about half the added nitrogen, was found in the final product. This lower nitrogen value indicates that structures of type XVII are present, and that the cross-linking of the lignin molecules has reduced the number of active sites available for reaction with the dimethylamine.

The concept of attaching pendant pesticidal fragments to a lignin backbone should also be capable of extension to include chemicals other than OMPA, and efforts were therefore made to synthesize lignin-phosphorus derivatives containing phosphate, thiophosphate, pyrophosphate, and dithiopyrophosphate groupings. These groupings are present in a large number of commercial insecticides, nematocides, and parasiticides.[66] The preparations of the lignin derivatives XVIII, XIX, XX, and XXI were therefore attempted.

$$
\begin{array}{cc}
OEt\quad\;\; OEt & OEt\quad\;\; OEt \\
|\quad\;\;\; / & |\quad\;\;\; / \\
lignin-O-P-O-P & lignin-O-P-O-P \\
\|\quad\;\; \|\backslash & \|\quad\;\; \|\backslash \\
O\quad\; O\;\; OEt & S\quad\; S\;\; OEt \\
XVIII & XIX
\end{array}
$$

$$
\begin{array}{cc}
\quad\;\; /NMe_2 & \quad\;\; /OEt \\
lignin-O-P & lignin-O-P \\
\|\backslash & \|\backslash \\
S\;\; NMe_2 & S\;\; OEt \\
XX & XXI
\end{array}
$$

Biological evaluation of all these lignin derivatives has yet to be carried out.

Interestingly, other phosphorus-containing lignin derivatives have been biologically evaluated by Russian researchers[67] and found to possess insecticidal activity. These are lignin thiophosphates prepared by the reaction of thiophosphoryl or diethylthiophosphoryl chlorides with acid hydrolysis lignins and their chloro- and nitro-derivatives. Since the polymeric derivatives alone cannot reasonably be regarded as biologically active, it must be assumed that either low molecular weight impurities are present, or that the high molecular weight polymer is slowly releasing insecticidal fragments. Similar work has been more recently carried out by Japanese investigators.[68, 69] In this case, the sodium salts of thiolignin and cellulose were condensed with dialkylphosphochloridothioate to yield pesticide-polymer combinations.

Analogously, some carboxymethyl ethers of kraft lignin have been shown to possess insecticidal[70] and fungicidal activity.[71] In this case, it seems more likely that the biological activity is due to the presence of low molecular weight aryloxyacetic acids. Nevertheless, the possibility of a controlled release mechanism operating to dispense biocides by breakdown of intralignin linkages cannot be completely excluded without further investigation.

A final example of the *ab initio* synthesis of a pesticide on a polymer is provided by the reaction of 2,3,4,5-tetrachloropyridine with lignin.[62] The product, which can be reasonably formulated as XXII, is a hydrolyzable lignin ether of 4-hydroxy-2,3,5-trichloropyridine. It is one of the family of very effective pyridine-based herbicides developed by the scientists of the Dow Chemical Company.

XXII

XI. FACTORS AFFECTING BIOCIDE LIBERATION

The release mechanism of all the aforementioned controlled release systems involves bond scission. Hydrolyzable linkages, of course, provide a cleavage mechanism which has the merit of being activated by the presence of water. Moisture also serves to initiate many natural phenomena, including the germination of seeds. Thus, a controlled dispensation of herbicides by hydrolysis is particularly apropos since it implies that the herbicide is likely to be released at just about the time when the embryonic weed is at its most susceptible stage. Furthermore, the most important herbicides are carboxylic acids which can readily and inexpensively be converted into polymeric esters.

It therefore follows that the release of the bioactive component from the grip of the polymer should be explainable by classical kinetic expressions describing hydrolytic reactions. However, although individually the hydrolysis reactions must follow first-order reaction kinetics, specific release rate constants will vary according to the bonding environment of the ester linkage. Under precise laboratory conditions, the rate of hydrolysis for an ester of a water insoluble polymer can be written as:

$$R_d = WdC/dt \qquad (14)$$

or by:

$$R_d = K_d WC \qquad (15)$$

where W is the amount of pesticide-polymer combination used, C is the concentration of the herbicide per unit weight contained therein at time t, and K_d is the degradation rate constant. When C_o is the value of C at time zero, then separation of the variables in Equation 14 and integration affords the relationship:

$$C = C_o e^{-K_d t} \qquad (16)$$

This equation implies that the rate at which the free herbicide is released merely depends upon the initial amount of material condensed onto the polymer in conjunction with the specific characteristics of the bond formed. Figure 4 presents data collected from hydrolysis experiments[72] carried out at various pH values on a series of ester combinations of 2,4-D on an α-cellulose backbone. Since it is obvious that the hydrolysis of α-cellulose-2,4-dichlorophenoxyacetate does not follow the predicted first order reaction kinetics scheme outlined above, the reasons for this apparent theoretical contradiction must be examined.

The esters studied in this project contained 4.1 or 21.5% by wt of combined herbicide. The data in Figure 4 shows that the release is dependent on both the degree of substitution and the pH of the hydrolysis medium. Moreover, since the hydrolysis samples contained the same quantity of combined active ingredient, it is clear that the fastest release of herbicide is obtained with the α-cellulose ester having the lowest degree of substitution. This pattern of herbicide release can be explained in terms of the microstructure of the α-cellulose fiber, which contains dispersed crystalline (50 to 70% by wt) and amorphous (50 to 30% by wt) regions.[73] Thus, the esterification of the herbicide does not take place on a uniform backbone. In fact, this occurrence would be virtually impossible with any naturally occurring polymer or polymer composite, such as bark. The amorphous region provides essentially all the readily available reactive hydroxy groups in nonpolar, nonswelling reaction media. In unswollen, native cellulose, those amount to only 0.4% of the total functionality.[74] This value is increased to 0.54% as a result of the isolation procedures, which generate about one third more hydroxy groups.[75] Thus, it can be anticipated that esterification reactions will begin randomly on accessible amorphous surfaces and subsequently spread along the polysaccharide chains. As a result, the density and pattern of substituents will vary throughout the α-cellulose substrate. This variance is compounded by the unequal activities towards esterification of the hydroxy groups on each anhydroglucose unit. For example, tosylation experiments suggest that the reactivities of the C_6, C_2, and C_3 hydroxy groups are in the ratio of 215:33:1, respectively.[76] Furthermore, a kinetic study demonstrated that the primary hydroxy group reacts 58 times faster than its secondary neighbors.[77]

As a result of all these facts, it can be concluded that the esterification reactions of α-cellulose at low degrees of substitution will randomly occur almost exclusively on

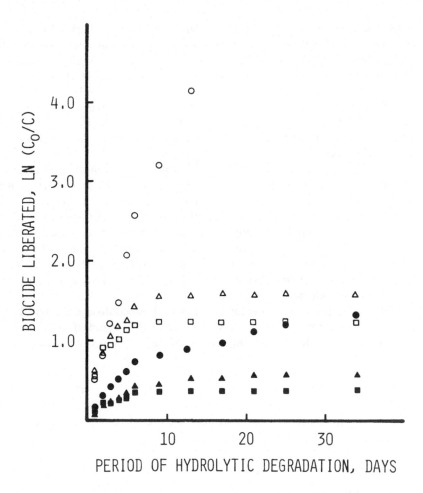

FIGURE 4. Characteristics of the in vitro hydrolysis of α-cellulose esters of 2,4-D containing 4.1% w/w (open symbols) or 21.5% w/w (solid symbols) combined herbicide when exposed to solutions of pH 4 (triangles), 7 (squares), and 10 (circles), respectively.

the fiber surfaces associated with the amorphous region at the primary hydroxy groups of the C_6 carbon atom of each glucose unit. It therefore follows that in the α-cellulose derivative there will be groups of more or less closely located ester units separated by areas having only an occasional ester substituent and, hence, a hydrophilic character comparable to that of the original amorphous region of the α-cellulose. Of course, as the degree of substitution is increased, the size of these hydrophilic areas must decrease. Ultimately, the hydrophilicity can become so small that water cannot effectively permeate the polymer matrix[78] to initiate hydrolytic release of the herbicide.[22, 23]

This structural picture therefore provides a satisfying explanation for the hydrolytic release in Figure 4, since each decrease in hydrophilicity will cause a decrease in the rate of hydrolysis. Obviously, a defining minimum limit must exist where further decreases in the amount of esterified hydrophobic groups will not lead to a faster rate of hydrolysis.

The release rate from all of these types of bioactive polymers is dependent upon the inherent properties of the combination, as well as the environmental conditions to which the controlled release combination is exposed. These results serve to explain earlier experiments in which simulated rainfall tests surprisingly revealed that combinations containing greater amounts of 2,4-D condensed with Douglas-fir sawdust or bark released less herbicide than did the corresponding combinations containing lower concentrations of the pendant herbicide.[79] It is also noteworthy that the release rates of 4-(2,4-dichlorophenoxy)butyric acid (2,4-DB) from bark combinations increased in the ratio 1:2:3 when western red alder (*Alnus rubra* Bong.), Douglas-fir, or hemlock (*Tsuga heterophylla*) were the source of the substrate, respectively. This reflects the different intrinsic chemistry and morphology of each of these barks. The fact that the release rate can also be modified by the selection of the polymeric substrate is also demonstrated by these results.

Modification of the polymeric backbone itself is another method used to alter the subsequent release of attached herbicide. Thus, modification of starch by cross-linking with epichlorohydrin or by reaction with acrylonitrile permits the subsequent preparation of 2,4-D starch esters with differing solubilities in water.[40] In aqueous hydrolysis tests (pH 6 to 8) the cyanoethylated controlled release herbicide released soluble ester and acid forms of 2,4-D at a slightly faster rate than did the unmodified starch formulation. In contrast, the epichlorohydrin cross-linked combination showed half the release rate of the unmodified starch derivative. These effects were also attributed to the relative hydrophobic nature of the esters.

In the soil, the behavior of the previously discussed α-cellulose ester will be more complex. At the critical time, t_c, which marks the end of the period of herbicidal effectiveness, the rate of release, R_d, for an application of weight W follows from Equation 16 so that

$$R_d = WC_o e^{-K_d t_c} \qquad (17)$$

Since the time to reach this critical point is actually the duration of herbicidal effectiveness (D_{pr}) then

$$D_{pr} = (1/K_d)(\ln W - \ln R_d/K_d) \qquad (18)$$

$$D_{pr} = M \log W - N \qquad (19)$$

where

$$M = 2.3/K_d$$

$$N = (2.3/K_d) \log (R_d/K_d)$$

Therefore, a plot of the duration of herbicidal effectiveness versus the logarithm of the weight of α cellulose-2,4-dichlorophenoxyacetate applied should give a straight line[23] intersecting the y-axis. This would indicate, by the magnitude of the intercept, the minimum amount ($W_m = 10^{N/M}$ when $D_{pr} = 0$) of the herbicide-polymer combination necessary for pesticidal effectiveness. Although the data in Figure 5 validates Equation 19, it must be remembered that the effectiveness of controlled release herbicides in the real world is less easily defined, since bioorganisms can alter the rate of hydrolysis. For example, the release of 2,4-D from an aqueous suspension of bark 2,4-dichlorophenoxyacetate occurs through hydrolytic cleavage, but in the soil this scission is supplemented by microbial and enzymic attack on the polymeric composite. However, it is known that rapid deterioration of cellulose occurs in soil,[41] whilst the physical properties of the monoacylated counterpart were not adversely affected by soil burial for 6 months.[80] Thus, a moderately bonded controlled release cellulose matrix need

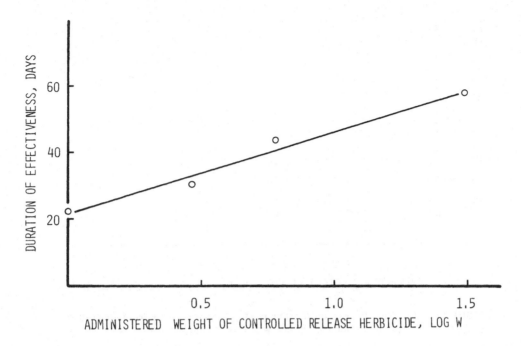

FIGURE 5. The dependence of the duration of effectiveness on the level of application for a chemical combination of a biocide with a water swellable, but insoluble, polymer.

not be destroyed within this short period of time because of the protective effect of the esterifying acid.

The potential contribution of the polymer backbone to the kinetics of herbicide release may be quite different if the polymer is water soluble.[23] This is particularly well exemplified by Neogi's studies of the duration of herbicidal effectiveness of water-soluble poly(vinyl alcohol) partially esterified (56.1% by wt.) with 4-chloro-2-methyl-phenoxyacetic acid. When such a combination in particulate form is saturated with water, the rate of hydrolytic degradation (R_h) for n spherical particles of radius r and density ϱ is given by

$$R_h = n\rho 4\pi r^2 \, dr/dt \tag{20}$$

and by

$$R_h = nK_h 4\pi r^2 C_o \tag{21}$$

where C_o is the initial concentration of pesticide-polymer linkages on the surface of a particle of radius r_o, and K_h is the hydrolysis rate constant. Because each biocide molecule liberated by hydrolysis exposes another biocide-polymer linkage beneath, C_o is a constant. The duration of pesticide release (D_{pr}) can then be obtained by combination of Equations 20 and 21 so that

$$D_{pr} = \rho r_o / K_h C_o - (R_h \rho^2 / n\pi K_h^3 C_o^3) \tag{22}$$

or

$$D_{pr} = A - BW^{-1/2} \tag{23}$$

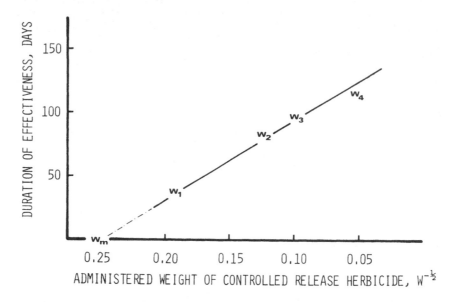

FIGURE 6.   The dependence of the duration of effectiveness on the level of application (W_{1-4}) for a chemical combination of a biocide with a water soluble polymer.

where A and B are constants, and W is the amount of the pesticide-polymer combinaton used. It is important to note that there is a minimum amount of the controlled release combination ($W_m = B^2/A^2$ when $D_{pr} = 0$) which must be applied to observe herbicidal activity (Figure 6).

XII. BIOEVALUATION OF CONTROLLED RELEASE SYSTEMS

Whatever the controlled release system ultimately employed, it is well to remember that in the field or forest the controlled release herbicide is neither regularly watered and maintained at a constant temperature, nor is it subjected to systematic challenges with carefully planted, premium quality seeds. In other words, predictions based upon elegant kinetic analysis and the subsequent results of tyrannically controlled laboratory experiments must be construed as shaky, to say the least. This is clearly illustrated by the long-term average climatological data for two field sites which have been intensively studied. The superimposition of the curves generated from average monthly rainfall and temperature data depicted in Figures 7[81] and 8[82] serve to illustrate the problems confronted in the outside world. These two limiting factors oppose each others influence upon the hydrolytic degradation of the lignocellulosic ester of 2,4-DB. Thus, the high temperatures of July and August in Washington, which would accelerate the rate of hydrolysis, are offset by the lack of water. Conversely, the abundant supply of water for hydrolysis in December and January is counterbalanced by the freezing temperatures during these months. Although there is much less rainfall in Mississippi, a similar general trend of rain and temperature opposition exists, with the exception of the month of July which characteristically experiences a limited period of heavy rainfall. Clearly then, indoor or outdoor pot studies can only provide a qualitative indication of the real-life activity of the bioactive polymer.

However, preliminary efficacy data can certainly be obtained by this means. Therefore, before field testing, three distinct 2,4-DB controlled release combinations with Douglas-fir bark were examined for their selective toxicity to deciduous brush in the presence of conifers.[83] The test plants selected were 2- to 3-year-old Douglas-fir and

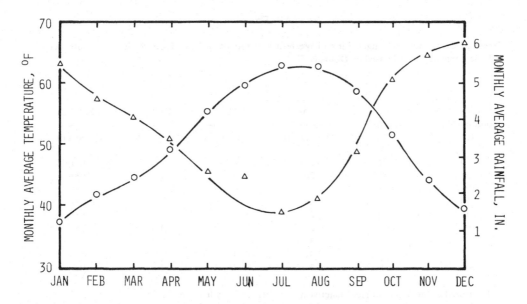

FIGURE 7. The climatological data for the Sedro Woolley field trial showing the general opposing trends of monthly average temperature (O) and rainfall (△).

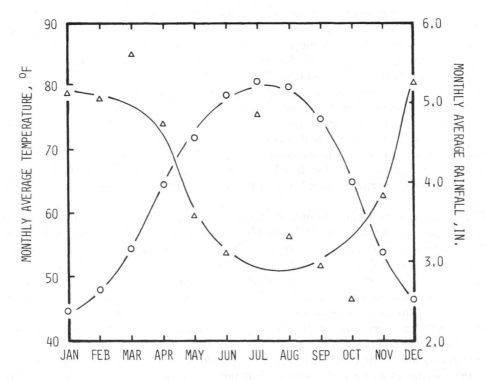

FIGURE 8. The climatological data for the Mississippi field trial showing the general opposing trends of monthly average temperature (O) and rainfall (△).

western red alder seedlings supplied by the Department of Natural Resources of the State of Washington. Each plant was individually transplanted outside in Seattle, Washington on December 6, 1969 into pots containing a weed-free soil having a surface area of 140 cm². The soil was an acidic (pH 6) Alderwood, sandy loam comprising

TABLE 8

Extent of Crown Kill in Douglas-Fir and Western Red Alder Seedlings Treated with the Controlled Release Form of 2,4-DB Based on Douglas-Fir Bark

Application of herbicide-bark combination (g)	Bioactivity (%)[a]	Average height of transplanted seedling at time of treatment (cm)		Extent of crown kill (%)	
		Douglas-fir	W. red alder	Douglas-fir	W. red alder
10	50	25	37	0	100
55	50	26	38	100	100
100	50	25	38	100	100
10	37	26	38	0	100
55	37	24	37	100	100
100	37	25	38	100	100
10	31	30	46	0	75
55	31	32	47	75	100
100	31	31	45	100	100

[a] Expressed as the weight of free bioactive agent liberatable by hydrolysis.

organic matter (15.2%) and a residuum of clay (15.1%), silt (22.7%) and sand (62.2%).

After the seedlings were well established, quadruplicate sets were treated in March 1970 by evenly distributing the polymeric form of 2,4-DB over the surface of each pot which was then sprinkled with more soil (150 g). The three herbicide-bark combinations were applied at levels of 10, 55, and 100 grams per pot and contained 31, 37, and 50% by wt of releasable herbicide.[17] Also, untreated, unplanted pots were maintained to measure the natural invasion of competitive vegetation.

The heights of all trees were measured at the time of transplanting and also in September 1970. The data is summarized in Table 8. The phytotoxic effect of the treatments was observed on individual seedlings at 15-day intervals and assessed using a percentage scale for the extent of crown damage. The values ranged from 10% for undamaged crowns to 100% for decedent crowns. This data was collected from each of the four replicates and averaged to give the results expressed as percent kill in Table 8.

The complete crown kill exhibited by the western red alder seedlings demonstrated that all of the polymeric forms of 2,4-dichlorophenoxy-γ-butyric acid are indeed biologically active at every level of application. The first signs of phytotoxicity appeared 25 days after the application of the controlled release combinations and the crowns were completely dead within a further 110 days. This phytotoxic action was also exhibited towards the Douglas-fir seedlings, but only at the higher levels of application where initial damage became apparent 45 days after treatment. Also, the realization of crown kill was more sluggish for the softwood seedlings, and 205 days elapsed before the seedlings expired.

Differentiation of the controlled release herbicide between the gymnosperm and angiosperm seedlings was greatly enhanced at the lowest level of application (10g). In fact, Douglas-fir seedlings were unharmed, while the western red alder seedlings succumbed. Of course, this implied that a herbicidal combination selectively phytotoxic to deciduous species was in hand. This conclusion was further supported by the measurements of seedling heights (Table 9), which showed that the controlled release combinations did not, in any way, adversely affect the growth of the Douglas-fir seedlings.

TABLE 9

Extent of Height Increase of Douglas-Fir and Western Red Alder Seedlings Treated with the Controlled Release Form of 2,4-DB Based on Douglas-Fir Bark, After One Growing Season

Application of herbicide-bark combination (g)	Bioactivity (%)[a]	Height Increase for			
		Douglas-fir (%)		W. red alder (%)	
		Untreated	Treated	Untreated	Treated
10	50	55	54	80	0
10	37	54	56	78	0
10	31	49	50	68	16

[a] Expressed as the weight of free bioactive agent liberatable by hydrolysis.

XIII. FIELD-SCALE TESTING OF CONTROLLED RELEASE HERBICIDES

It has already been emphasized that the only true test of the efficacy of any controlled release system is the actual result achieved in the field. The first field trial using the bark-2,4-DB ester was established at the site named "Muddy Creek" near the town of Sedro Woolley, Washington in cooperation with the Department of Natural Resources of the State of Washington.

Muddy Creek was logged in the winter of 1969 and burned during August 1970 without significant damage to the humus layer. The cleared ground was replanted with 2- to 3-year-old Douglas-fir seedlings by the Department of Natural Resources during December 1970 and January 1971, using stock from their own nursery.

Twenty rectangular plots within the site were marked off in March 1971. Each plot was about 500 m² in area and contained between 32 to 40 Douglas-fir seedlings. The controlled release combination of 2,4-DB and Douglas-fir bark contained 37% by wt of releasable herbicide and was custom synthesized by the Eastman Kodak Company (Rochester, N.Y.).

The levels of application to a square area (2.32 m²) around each seedling were 0, 32, 80, 200, and 320 g a.i./m². An equivalent area around other untreated seedlings was initially hand-weeded and covered with 5-mil-thick black agricultural-grade polypropylene film (N300) supplied by Hercules, Inc. of Wilmington, Del. The heights of all seedlings were recorded after treatment in April 1971 and subsequently monitored in each succeeding September at the end of the growing season. The sixth-year growth data for each of the treatments is summarized in Tables 10 and 11. Even a cursory examination of only the arithmetic means of the cumulative growth discloses the obvious differences which are attributable to the application of the controlled release herbicide to the conifers. These differences are more rigorously defined by expressing conifer growth as a function of time (Figure 9). This data unequivocally shows that the conifers which received the controlled release herbicide treatment in 1971 have grown significantly faster than the trees which were not treated. In fact, the difference in growth amounts to 15% in the best case. This growth enhancement effect cannot be entirely attributed to the elimination of competitive vegetation around the conifers because trees mechanically maintained weed-free by the surrounding plastic film grew at a rate which is about 10% less than their counterparts which received the controlled

TABLE 10

Statistical Evaluation of the Growth of Douglas-Fir Seedlings at the End of the Fifth Year[a]

Application of herbicide-bark combination (g)	Mean growth of seedlings from Sept. 1974 to Sept. 1975 (cm)				Mean growth of seedlings from March 1971 to Sept. 1975 (cm)				Height as of Sept. 1975 (cm)	
	Arithmetic mean	Standard deviation	Arithmetic mean (dbh)[b]	Standard deviation	Arithmetic mean	Standard deviation	Arithmetic mean (dbh)	Standard deviation	Arithmetic mean	Standard deviation
0	93	17	1.8	0.4	307	47	3.9	0.9	348	50
200	91	21	1.9	0.4	355	58	4.6	1.1	392	61
500	98	21	2.1	0.4	361	30	4.9	0.9	404	31
1250	95	21	2.2	0.5	348	36	5.0	1.0	396	41
2000	90	16	2.3	0.7	321	44	4.6	1.1	368	44
zero[c]	95	20	2.2	0.5	343	40	4.6	0.9	378	44

[a] Growth data of Sedro Woolley field trial as of September 1975.

[b] Seedling diameter at breast height (137 cm).

[c] Seedlings surrounded by black polypropylene film (2.26 m^2).

TABLE 11

Statistical Evaluation of the Growth of Douglas-Fir Seedlings at the End of the Sixth Year[a]

Application of herbicide-bark combination (g)	Mean growth of seedlings from Sept. 1975 to Sept. 1976 (cm)		Mean growth of seedlings from March 1971 to Sept. 1976 (cm)				Height as of September 1976 (cm)	
	Arithmetic mean (dbh)[b]	Standard deviation	Arithmetic mean	Standard deviation	Arithmetic mean (dbh)	Standard deviation	Arithmetic mean	Standard deviation
0	0.9	0.3	432	63	4.7	0.9	472	66
200	1.0	0.3	486	81	5.5	1.1	528	82
500	1.2	0.6	501	48	6.0	0.8	544	48
1250	0.8	0.5	465	63	5.9	0.9	513	69
2000	0.8	0.3	450	64	5.4	0.9	499	65
zero[c]	0.8	0.3	460	55	5.4	0.9	498	56

Note: The first data column group also includes arithmetic mean values of 124, 136, 140, 118, 131, 121 and standard deviation values of 30, 37, 25, 44, 30, 29 for rows 0, 200, 500, 1250, 2000, zero respectively.

a Growth data of Sedro Woolley field trial as of September 1976.

b Seedling diameter at breast height (137 cm).

c Seedlings surrounded by black polypropylene film (2.26 m²).

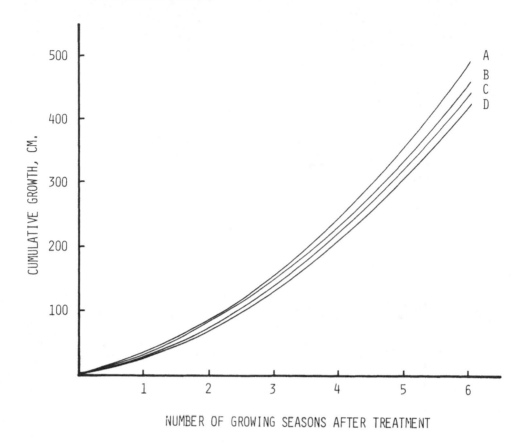

FIGURE 9.   Plots of regression equations for the cumulative growth of Douglas-fir seedlings at the Sedro Woolley forest plantation. A, data from the application of the 80 g a.i./m² bark ester of 2,4-DB; B, data from the placement of the polypropylene film; C, data from the 320 g a.i./m² application of the bark ester of 2,4-DB; D, data from the untreated controls.

release herbicide applications. However, these plastic-surrounded trees did grow about 6% faster than the untreated trees. Presumably, this value represents the maximum improvement in growth that can be expected from complete weed control.

Comparison with the extensive growth history curves in Figure 10 demonstrates that both the controlled release herbicide and plastic film treated trees are growing faster than any Douglas-firs previously on record. Clearly, the superior growth observed as a result of application of the controlled release combination suggests that the release of the auxin-like herbicide is providing the conifer with a metabolic stimulus. Of course, this is in addition to the several benefits accruing from the suppression of competitive vegetation. An annual plot (Figure 11) of the cumulative yearly growth as a function of the level of controlled release herbicide applied supports this hypothesis. The emergence of a well-defined maximum in the conifer growth is evident from the data presented. That is, the higher application level of the controlled release herbicide does not provide the same growth enhancement as the lower levels. Obviously, the higher levels cannot be less effective than the lower in the suppression of competitive vegetation. Therefore, it can be concluded that the level of herbicide appropriate for the metabolic stimulation of the conifer has been exceeded and that the level of phytotoxicity is being approached. The benefits from the early metabolic stimulation are still accruing, since the cumulative conifer growth induced by all but the highest controlled release herbicide applications are statistically different from the growth of the

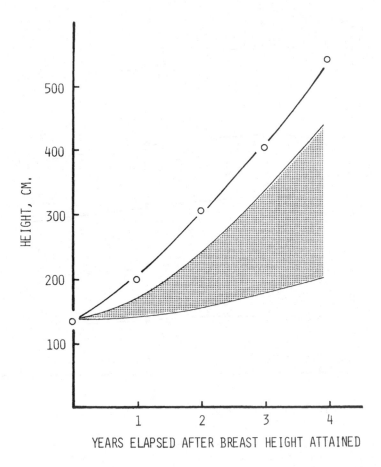

FIGURE 10. Douglas-fir growth history curves of the Pacific Northwest (hatched area) and of the most effective controlled release treatment (O) at Sedro Woolley.

untreated trees from 1971 through 1976. However, only the application level of 500g a.i./m² of the controlled release herbicide is still inducing a yearly growth acceleration relative to the controls in 1976.

The merit of this particular application level of the controlled release herbicide is further demonstrated by comparison of the trunk cross-sectional areas 4.5 ft above ground level (dbh). The comparative data in Table 12 shows that the most effective controlled release herbicide treatment has increased the trunk cross-sectional area by 63% relative to the untreated controls. Once again, the growth stimulatory effect of the metabolic herbicide is demonstrated by the fact that the cross-sectional area of the trees freed from competition by the surrounding polypropylene film was only 33% greater than the untreated controls.

The potential economic importance of these results is emphasized by transformation of the trunk cross-sectional areas into wood volumes (Table 13). The preferred controlled release herbicide treatment affords a wood volume increase of 40, 54, and 74% in years 4, 5, and 6, respectively, relative to the volume of wood produced by the trees grown in the conventional way. Also, it is noteworthy that after the sixth year of growth, the presence of the controlled release herbicide contiguous to the growing tree can induce the growth of a 27% greater amount of wood than that produced by a tree mechanically maintained free of competitive vegetation. In other words, these results

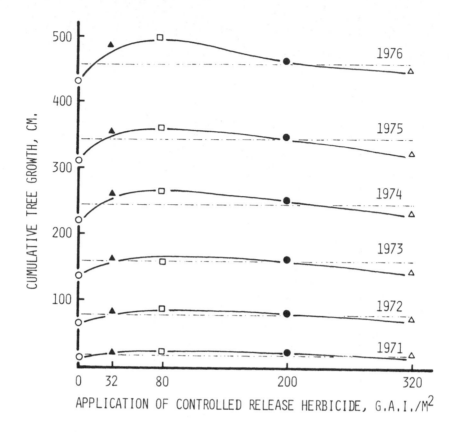

FIGURE 11.   The cumulative yearly growth of Pacific Northwest trees receiving different amounts of the controlled release herbicide where (-··-··-) represents the cumulative growth of the trees surrounded by the polypropylene film.

show that not only are the trees receiving the controlled release herbicide treatment taller than their untreated counterparts, they are photosynthetically more effective. Although the technical merit of the use of controlled release herbicides in reforestation is clearly established by this long-term experiment, a number of important questions remain to be answered. For instance, the preferred size of the circumjacent area of treatment has not yet been defined, nor have the application rates for conifers been optimized for other areas within and without the United States.

The second major field trial designed to test the utility of the bark-2,4-DB ester was installed on the site known as Electric Mills near Starkville, Miss. in the winter of 1973. In this experiment, the treated area around each seedling was reduced to 1.49 m², and one-year-old loblolly pine (*Pinus taeda L.*) seedlings were planted instead of the Douglas-fir used in the Pacific Northwest. This planting took place immediately following the conventional site preparation procedures of a logged area in the southern United States. The 2,4-DB combination was part of the batch originally synthesized for the Sedro Woolley test which contained 37% by wt of releasable herbicide. Equal amounts of free herbicide were also tested for comparison with the bioactive polymer combination. All of the treatments were tested at two different rates (6.2 and 24.8 g a.i./m²). The highest amount of active ingredient which was applied to the area around the young stems is, therefore, 7 g/m² less than the lowest treatment rate at the Muddy Creek field trial. That is, not only was the treated area per seedling reduced, but the application rate was similarly diminished. Furthermore, one area was repetitively

TABLE 12

Growth and Structural Analysis of Douglas-Fir Seedlings at Sedro Woolley

		Trunk cross-sectional area at breast height for seedlings (cm^2)				
Year data taken[a]	Untreated	Treated with herbicide-bark combination (g)				Surrounded by plastic film[b]
		200	500	1250	2000	
1974	3.40	5.90	6.03	6.61	4.56	4.56
1975	12.01	16.40	18.70	19.87	16.76	16.98
1976	17.50	23.67	28.46	27.06	22.56	23.24

[a] All data was taken in September of the year.
[b] Seedlings surrounded by black polypropylene film (2.26 m^2).

TABLE 13

Growth and Structural Analysis of Douglas-Fir Seedlings at Sedro Woolley

		Trunk board-foot volume of seedlings[b]				
Year data taken[a]	Untreated	Treated with herbicide-bark combinaton (g)				Surrounded by plastic film[c]
		200	500	1250	2000	
1974	0.60	0.84	0.84	0.96	0.72	0.72
1975	1.56	2.16	2.40	2.64	2.16	2.28
1976	2.28	3.24	3.96	3.60	3.00	3.12

[a] All data was taken in September of the year.
[b] Calculated using the assumption that the trunk is a perfect cone.
[c] Seedlings surrounded by black polypropylene film (2.26 m^2).

hand-weeded, while five others were designated as untreated controls. Finally, the heights of all seedlings were initially recorded in March of 1973 and subsequently monitored at the end of each succeeding growing season.

The results collected 3 years after the initiation of the experiment (Figure 12) show a growth enhancement of the loblolly pine seedlings treated with the combination of herbicide and biodegradable substrate. The growth caused by this treatment exceeded that of the seedlings which had received no treatment, as well as the growth of those trees which had been exposed to quantities of the free herbicide corresponding to the amounts potentially available from the application of the controlled release combination. However, the most obvious initial benefit from the use of the latter is the number of seedlings surviving. These values are 11 to 51% greater than when the conventional biocide (Figure 13) was used. Neither of the bioactive-polymer applications induced any significant survival differences when compared with the untreated control seedlings. It is noteworthy that the controlled release formulation does not become toxic in the second year when survival was independent of treatment or the application rate.

The dimensions of the seedlings receiving the low and high application levels of the controlled release herbicide surpassed those of the untreated controls by 17 and 23% in height and 30 and 52% in diameter, respectively, at the end of three years. Concomitantly, the seedlings which received the higher bioactive-polymer application exceeded

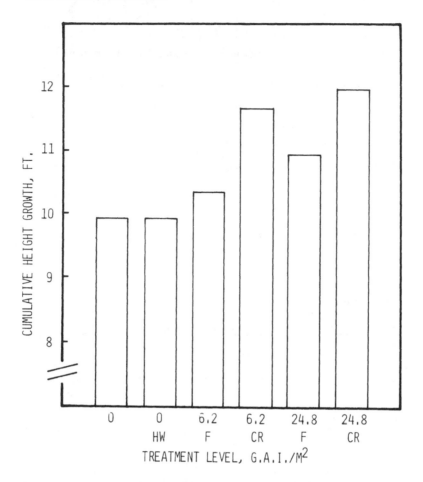

FIGURE 12.   The cumulative height growth of herbicide-treated loblolly pine seed-lings (1 - 0) 3 years after field outplanting in Mississippi. F, free 2,4-DB; CR, controlled release combination; HW, hand-weeded controls.

those which had been treated with the free herbicide itself by 16% in height and 30% in diameter at breast height. The lower application rate of the controlled release herbicide stimulated a 14% height and a 42% diameter augmentation in comparison to the young pines which received an application of the same rate of the free herbicidal acid.

The trunk cross-sectional areas and the wood volume of the loblolly pine seedlings 3 years after the experiment was established are presented in Figures 14 and 15, respectively. As in the Muddy Creek field trial, the growth enhancement effects noticeable in the Electric Mills trial are attributable to the action of the controlled release herbicide. It is noteworthy that there is an increase of 69 and 64% in trunk cross-sectional area and volume, respectively, as a result of the lower level of application of the controlled release herbicide. The best, but not necessarily optimum, application rate exceeded even these values when similarly compared to the untreated controls. The cross-sectional area at breast height and wood volume are respectively 131 and 125% greater than those of trees left untreated. Although slight growth increases are visible with the application of the conventional herbicide, the excessive mortality stemming from the presence of the free material precludes this mode of operation. That is, the net accumulation of wood is vastly decreased since there are far fewer surviving seedlings.

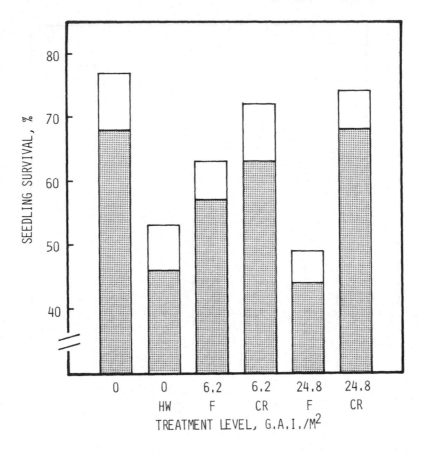

FIGURE 13.   The cumulative survival of herbicide-treated loblolly pine seedlings
(1 - 0) 1 (hatched area) and 2 (total bar) years after field outplanting in Mississippi.
F, free 2,4-DB; CR, controlled release combination; HW, hand-weeded controls.

Finally, the growth enhancement of those trees which were repetitively handweeded is negligible, suggesting that a metabolic stimulus, rather than the elimination of competitive vegetation, is responsible for the growth augmentation observed in Mississippi.

All of these results indicate that the biodegradable controlled release herbicide combination is at least as effective on loblolly pine grown in the southern part of the U.S. as it has been shown to be on Douglas-fir in the Pacific Northwest, in spite of the difference in climatological environments.

The general adoption of controlled release herbicides into U.S. operational conifer-reforestation practices would, therefore, seem to be only a matter of time, since the crucial underlying economics seem to be highly favorable in all cases thus far examined.[84]

XIV. CONCLUSION

It is apparent that even within the limited technological area covered in this chapter, many alternatives are open to the designer of practical and useful controlled release delivery systems.

Any selection from among these will always be a difficult task requiring a variety of ecological, scientific, and economic inputs.

Based on a background of some 70 man-years of research on controlled release sys-

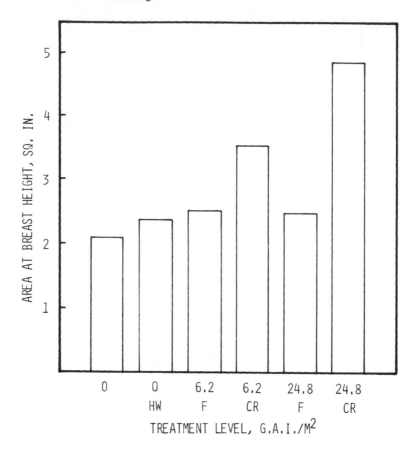

FIGURE 14.   The cross-sectional areaa at breast height of herbicide-treated loblolly pine seedlings (1 - 0) 3 years after field outplanting in Mississippi. F, free 2,4-
DB. CR, controlled release combination; HW, hand-weeded controls.

tems of all types, it is the firm belief of the University of Washington team that combinations of biocides with natural polymers will generally offer one of the best ap-
proaches to developing a specific pesticide-delivery mechanism which will be cost-
effective and noninjurious to the environment.

ACKNOWLEDGMENTS

For outstanding logistic support in forestry over the years thanks are due to D. L.
Fraser, D. R. Hopkins, L. V. Morton, W. A. Nelson, H. W. Anderson, D. C. Belz,
S. K. Thompson, K. L. Reed, and J. M. Finnis of the Department of Natural Re-
sources of the State of Washington (Bert L. Cole, Commissioner of Public Lands).
Forestry cooperation in Mississippi was generously provided by Professor W. F. Miller
and R. C. Maggio of Mississippi State University.

The contributions of the following group of students and faculty from the U. S.
and elsewhere who have worked on controlled release systems at the University of
Washington are gratefully acknowledged: T. J. Ahern, K. Akagane (Japan), D. Alker
(England), F. J. Allan III, G. G. Allan II, I. M. Allan, L. A. Allan, R. G. Allan, W.
Balaba (Uganda), D. A. Briggs, M. W. Byron, C. S. Chopra (India), C. Clinton, G.
D. Crosby, H. Dutkiewicz (Poland), J. Dutkiewicz (Poland), N. C. Foster, J. R. Fox,
J. F. Friedhoff, E. J. Gilmartin, P. Graef, M. K. Graef, D. D. Halabisky (Canada),

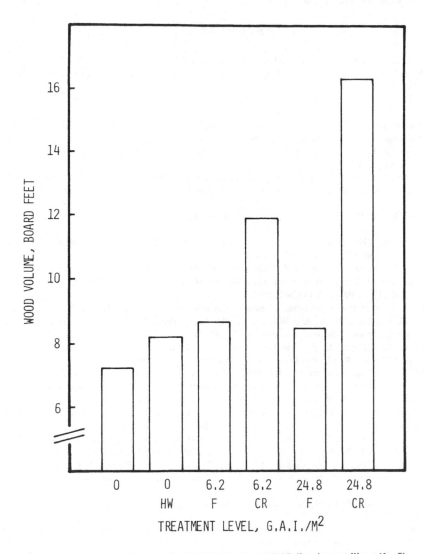

FIGURE 15.   The wood volume of herbicide-treated loblolly pine seedlings (1 - 0)
3 years after field outplanting in Mississippi. F, free 2,4-DB. CR, controlled release
combination; HW, hand-weeded controls.

D. M. Hanson, L. Hartikainen (Finland), C. W. Ingalls, D. Jackson (England), D.
M. Kirk, Y. Ko (S. Korea), J. Krumpos, J. Lee (Taiwan), W. W. Lee (Hong Kong),
T. Liu (Taiwan), D. W. MacGregor, M. Maggi, W. J. McConnell, T. Mehary (Ethio-
pia), L. E. Merrifield, J. M. Miller (England), M. L. Miller, A. N. Neogi (India), S.
Neogi (India), J. Palicka (Sweden), T. Popoff (Sweden), E. M. Passot (France), J. C.
Powell, S. C. Roberts, R. M. Russell, P. Rygiewicz, M. So (Hong Kong), S. So (Hong
Kong), J. D. Somers, H. Struszczyk (Poland), N. Thao (Laos), S. Thao (Laos), M.
L. Thomas, S. Venkataraman (India), R. M. Wilkins (England), P. L. Winslow, D.
Work and A. Yahiaoui (Algeria).

REFERENCES

1. Rice, E. L., *Allelopathy,* Academic Press, New York, 1974.
2. del Moral, R. and Müller, C. H., Allelopathic effects of *Eucalyptus camaldulensis, Am. Midl. Nat.* 83, 254, 1970.
3. Whittaker, R. H. and Feeny, P. P., Allelochemics: chemical interactions between species, *Science,* 171, 757, 1971.
4. Whittaker, R. H., The biochemical ecology of high plants, in *Chemical Ecology,* Sondheimer, E. and Simeone, J. B., Eds., Academic Press, New York, 1970, 43.
5. Goulding, R. L., Ed., *Proc. 1977 Int. Controlled Release Pesticide Symp.,* Oregon State University, Corvallis, Ore., 1977.
6. Kaufman, D. D. and Kearney, P. C., Microbial transformations in the soil, in *Herbicides, Physiology, Biochemistry, Ecology,* Vol. 2, 2nd ed., Audus, L. J., Ed., Academic Press, New York, 1976, chap. 2.
7. Ashbee, K. H. G., Water damage in glass fiber/polyester resin composites, *Chem. Abstr.,* 77, 152982z, 1977.
8. Crafts, A. S., *The Chemistry and Mode of Action of Herbicides,* Interscience, New York, 1961, 217.
9. Faerber, G., British Patent 826,831, 1960.
10. Pittman, C. U., Stahl, G. A., and Winters, H., Synthesis and mildew resistance of vinyl acetate and ethyl acrylate films containing chemically anchored fungicides, *J. Coat. Technol.,* 50, 49, 1978.
11. Baltazzi, E., U. S. Patent 3,343,941, 1967.
12. Flory, P. J., *Principles of Polymer Chemistry,* Cornell University Press, Ithaca, N. Y., 1953, 347.
13. Allan, G. G., Gara, R. I., and Wilkins, R. M., Controlled release pesticides. V. Phytotoxicity of some systemic insecticides to Spanish cedar, *Int. Pest Control,* 15(1), 4, 1973.
14. Allan, G. G., Chopra, C. S., Friedhoff, J. F., Gara, R. I., Maggi, M. W., Neogi, A. N., Roberts, S. C., and Wilkins, R. M., Pesticides, pollution and polymers, *Chem. Technol.,* 3, 171, 1973.
15. Allan, G. G., Belgian Patent 706,509, 1966; Australian Patent 423,990, 1967; Italian Patent 820,813, 1967; French Patent 1,544,406, 1968; Japanese Patent 558,788, 1969; U.K. Patent 1,212,842, 1970; Canadian Patents 855,181, 1970, 863,310, 1971; U.S. Patent 3,813,236, 1976.
16. Akagane, K., and Allan, G. G., Studies on long term protection. IV. Antifouling polymers for controlling sea weeds, *Shikizai Kyokaishi,* 46, 370, 1973.
17. Allan, G. G., Chopra, C. S., Neogi, A. N., and Wilkins, R. M., Controlled release pesticides. II. The synthesis of herbicide-forest solid waste combinations *Tappi,* 54, 1293, 1971.
18. Allan, G. G., Beer, J. W., Cousin, M. J., McConnell, W. J., Powell, J. C., and Yahiaoui, A. Polymeric drugs for plants, in *Polymeric Drugs,* Donaruma, L. G. and Vogl, O., Eds., Academic Press, New York, 1978, 193.
19. Beer, J. W., Controlled Release Herbicides in Reforestation, M.S. thesis, University of Washington, Seattle, 1978.
20. Cousin, M. J., The Controlled Release of Herbicides from Biodegradable Substrates, Ph. D. thesis, University of Washington, Seattle, 1978.
21. Friedhoff, J. F., The Potential of Controlled Release Herbicides for the Suppression of Unwanted Vegetation, Ph.D. thesis, University of Washington, Seattle, 1975.
22. Wilkins, R. M., The Design of Polymers for the Sustained Release of Selected Herbicides, M.S. thesis, University of Washington, Seattle, 1969.
23. Neogi, A. N., Polymer Selection for Controlled Release Pesticides, Ph.D. thesis, University of Washington, Seattle, 1970.
24. Jakubka, H. D. and Busch, E., Synthese von kovalent an polymere Trager gebundenen niedermolekularen Wirkstoffen und Versuche zur enzymatischen Reaktivierung, *Z. Chem.,* 13, 105, 1973.
25. Allan, G. G., Fox, J. R., and Kong, N. A., Critical Evaluation of the Potential Sources of Chitin and Chitosan, in Proc. 1st Int. Conf. on Chitin and Chitosan, Boston, MIT Sea Grant Program, Massachusetts Institute of Technology, Cambridge, Mass., 1978.
26. Naruse, H. and Maekawa, K., Controlled release pesticides. I. Dithiocarbamate derivatives, *J. Fac. Agr., Kyushu Univ.,* 21, 107, 1977.
27. Naruse, H. and Maekawa, K., Controlled release pesticides — part 2: Fungicidally active water soluble polymer, *J. Fac. Agric. Kyushu Univ.,* 21, 153, 1977.
28. Beasley, M. L. and Collins, R. L., Water-degradable polymers for controlled release of herbicides and other agents, *Science,* 169, 769, 1970.
29. Allan, G. G., Chopra, C. S., Neogi, A. N., and Wilkins, R. M., Degradable pesticides, *Chem. Br.,* 7(3), 129, 1971.
30. Smith, R. M. and Martell, A. E., *Critical Stability Constants,* Vol. 4, Plenum Press, New York, 1976.

31. **Neogi, A. N. and Allan, G. G.**, Controlled release pesticides: concepts and realization, in *Controlled Release of Biologically Active Agents*, Tanquary, A. C. and Lacey, R. E., Eds., Plenum Press, New York, 1974, 195.

32. **Allan, G. G.,**unpublished data, 1964.

33. **Dovlatyan, V. V. and Kostanyan, D. A.**, Polyvinyl esters of aryloxyacetic acids, *Izv. Akad. Nauk Arm. SSR Khim. Nauki*, 18, 325, 1965; as cited in *Chem. Abstr.*, 64, 623h, 1966.

34. **Feld, W. A., Post, L. K., and Harris, F. W.**, Controlled release from polymers containing pesticides as pendant substituents, in *Proc. 1975 Int. Controlled Release Pesticide Symp.*, Harris, F. W., Ed., Wright State University, Dayton, Ohio, 1975, 113.

35. **Allan, G. G. and Neogi, A. N.**, U. S. Patent 4062, 855, 1977.

36. **Akagane, K. and Allan G. G.**, Studies on long term protection. VI. Antifouling properties of tributyltin polymeric pesticides, *Shikizai Kyokaishi*, 46, 622, 1973.

37. **Akagane, K. and Allan, G. G.**, Studies on long term protection. III. Effect of main chain degradable polymeric pesticides to long term protection, *Shikizai Kyokaishi*, 46, 7,1973.

38. **Allan, G. G.,**Canadian Patent 855,181, 1970.

39. **Allan, G. G. and Halabisky, D. D.**, Fibre surface modification. XI. The covalent attachment of fungicides to lignocellulosic fibers, *J. Appl. Chem. Biotechnol.*, 21, 190, 1971.

40. **Mehltretter, C. L., Roth, W. B., Weakley, F. B., McGuire, T. A., and Russell, C. R.**, Potential controlled release herbicides from 2,4-D esters of starches, *Weed Sci.*, 22, 415, 1974.

41. **Allan, F. J., Allan, G. G., and Crank, G.**, Fungitoxicity of rhodanine derivatives, *J. Appl. Chem.*, 11, 68, 1961.

42. **Allan, F. J., Allan, G. G., Crank, G., and Thomson, J. B.**, Fungitoxicity of rhodanine derivatives, *J. Appl. Chem.*, 12, 46, 1962.

43. **Neumeyer, J., Gibbons, D., and Trask, H.**, Pesticides. I, *Chem. Week*, 104(15), 37, 1969.

44. **Allan, G. G., Akagane, K., Neogi, A. N., Reif, W. M., and Mattila, T.**, Physical entrapment of polyelectrolytes within microporous solids: the "Jack-in-the-box" effect, *Nature (London)*, 225, 175, 1970.

45. **Allan, G. G., Laine, J. E., and Neogi, A. N.**, Surface properties in relation to the bonding of nonwovens, in *Surface Characteristics of Fibers and Textiles, Part II,* Schick, M. J., Ed., Marcel Dekker, New York, 1977, chap. 16.

46. **Akagane, K. and Allan, G. G.**, Studies on long term protection. II. The effects of fungicidal polymers to long term protection, *Shikizai Kyokaishi*, 45, 479, 1972.

47. **Nagayoshi, A.**, personal communication, 1975.

48. **McCormick, C. L., Savage, K. E., and Hutchinson, B.**, Development of controlled-release polymer systems containing pendant metribuzin, in *Proc. 1977 Int. Controlled Release Pesticide Symp.*, Goulding, R. L., Ed., Oregon State University, Corvallis, Ore., 1977, 28.

49. **Schacht, E. H., Desmarets, G. E., Goethals, E. J., and St. Pierre, T.**, Synthesis and characterization of 2,6-dichlorobenzaldehyde-generating polymers, in *Polymeric Drugs*, Donaruma, L. G. and Vogl, O., Eds., Academic Press, New York, 1978, 331.

50. **Hilton, J. L.**, Regulation of mode of action of herbicides to controlled release technology, in *Proc. 1975 Int. Controlled Release Pesticide Symp.*, Harris, F. W., Ed., Wright State University, Dayton, Ohio, 1975, 303.

51. **Bittner, S., Perry, I., and Knobler, Y.**, Sustained release of cytokinins from natural polymers, *Phytochemistry*, 16, 305, 1977.

52. **Akagane, K. and Matsuura, K.**, Studies on long term protection. I. Effects of toxic polymers on long term protection, *Shikizai Kyokaishi*, 45, 69, 1972.

53. **Allan, G. G., Chopra, C. S., and Mattila, T.**, Imidazoles. III. Pesticidal screening of some substituted 1-hydroxyimidazoles, *Pestic. Sci.*, 3(2), 153, 1972.

54. **Akagane, K., Allan, G. G., and Passot, E. M.**, Fiber surface modification. XVII. The stereotop chemistry of the adsorption of polyaza heterocycles onto cellulose, *Pap. Puu.*, 56, 1, 1974.

55. **Akagane, K. and Allan, G. G.**, Studies on long term protection — part 5: Mutual interactions of active toxicants in antifouling polymers consisting of multicomponent toxicants, *Shikizai Kyokaishi*, 46, 437, 1973.

56. **Akagane, K.,**Japanese Patent 74,125,521, 1974.

57. **Allan, G. G., Killingbeck, J. E., and Roberts, S. C.**, Fiber surface modification — part 18: Kinetics and optimization of the reactions of cyanuric chloride with groundwood fibers and kraft lignin, *Pap. Puu*, 56, 945, 1974.

58. **Martin, H.**, Factors affecting the practical employment of systemic insecticides, *Can. Entomol.*, 82, 247, 1950.

59. **Spencer, E. Y. and O'Brien, R. D.**, Schradan; enhancement of anticholineterase activity in octamethylpyrophosphoramide by chlorine, *J. Agric. Food Chem.*, 1, 716, 1953.

60. **Kilby, B. A.**, The biochemistry of schradan, *Chem. Ind. (N.Y.)*, 856, 1953.
61. **Muhlmann, R. and Schrader, G.**, Hydrolyze der insektiziden Phosphorsäureester, *Z. Naturforsch.*, 12b, 196, 1957.
62. **Allan, G. G.**, unpublished data, 1963.
63. **Brown, D. M.**, *Advances in Organic Chemistry*, Vol. 3, Interscience, New York, 1960, 88.
64. **Gruze, H. and Koransky, W.**, Phosphorylation of nucleosides with dichlorophosphoric anhydride, *Angew. Chem.*, 71, 407, 1959.
65. **van Wazer, J. R.**, *Phosphorus and Its Compounds*, Interscience, New York, 1958, 247.
66. **Johnson, O.**, Pesticides '72. *Chem. Week*, 110(25), 33, 1972; Pesticides '72. II, *Chem. Week*, 111(4), 17, 1972.
67. **Tronov, B. V., Pershina, L. A. Morozova, V. M., Kovalenko, A. V., and Galochkin A. I.**, Thiophosphate products from hydrolytic lignin and their insecticidal properties, *Gidroliz. Lesokhim. Prom.*, 14(5), 10, 1961; as cited in *Chem. Abstr.*, 63, 1801g, 1962.
68. **Lee, C. and Maekawa, K.**, Pesticides combined with polymeric substances. I. Synthesis of cellulose-6-(diethyl phosphorothioate), *J. Fac. Agric. Kyushu Univ.*, 19, 1, 1974.
69. **Lee, C. and Maekawa, K.**, Pesticides combined with polymeric substances. II. Synthesis of thiolignin-(dialkyl phosphorothioate), *J. Fac. Agric. Kyushu Univ.*, 19, 65, 1974.
70. **Nikitin, V. M. and Kroshilova, T. M.**, Phenoxycarboxylic acids of lignin, *Izv. Vyssh. Uchebn. Zaved. Lesn. Zh.*, 7, 149, 1964; as cited in *Chem. Abstr.*, 62, 10660e, 1965.
71. **Ali, M. E., Ahmed, Q. A., and Khundkar, M. H.**, Evaluation of fungicidal properties of lignin acetic acid. I. Studies on fungus organisms in petri-plates, *Pak. J. Sci. Ind. Res.*, 1, 79, 1958.
72. **Allan, G. G., Beer, J. W., and Cousin, M. J.**, Controlled release of herbicides from biodegradable substrates, in *Controlled Release Pesticides*, Scher, H. B., Ed., American Chemical Society, Washington, D.C., 1977, 94.
73. **Browning, B. L.**, *The Chemistry of Wood*, Interscience, New York, 1963, 137.
74. **Timell, T.**, Studies on Cellulose Reactions, Ph.D. thesis, Royal Institute of Technology, Stockholm, Sweden, 1960, 23.
75. **Jorgensen, L. and Ribi, E.**, X-ray analysis of the swelling of wood pulp in sodium hydroxide solution, *Nature (London)*, 166, 148, 1950.
76. **Gardner, T. S. and Purves, C. B.**, The distribution of acetyl groups in a technical acetone-soluble cellulose acetate, *J. Am. Chem. Soc.*, 64, 1539, 1942.
77. **Honeyman, J.**, Reactions of cellulose. I, *J. Chem. Soc.*, 168, 1947.
78. **Higuchi, T. and Aguiar, A.**, A study of permeability to water vapor of fats, waxes and other enteric coating materials, *J. Pharm. Sci.*, 48, 577, 1959.
79. **Allan, G. G.**, U.S. Patent 3,813,236, 1974.
80. **Buras, E. M., Jr., Cooper, A. S., Keating, E. J., and Goldthwait, C. F.**, Practical partial acetylation of cotton, *Am. Dyest. Rep.*, 43, 205, 1954.
81. **Anon.**, Climatological data: Washington, National Climatic Center, Asheville, N.C., U.S. Department of Commerce, 79(13), 1, 1975.
82. **Anon.**, Climatological data: Mississippi, National Climatic Center, Asheville, N.C., U.S. Department of Commerce, 80(13), 1, 1975.
83. **Allan, G. G., Chopra, C. S., and Russell, R. M.**, Controlled release pesticides. III. Selective suppression of weeds and deciduous brush in the presence of conifers, *Int. Pest Control*, 14(2), 15, 1972.
84. **Allan, G. G., Powell, J. C., Cousin, M. J., McConnell, W. J., and Yahiaoui, A.**, Stimulation of timber growth by controlled release herbicides — opportunities and economic considerations in the Pacific Northwest, in *Chemical Marketing and Economics Reprints, Economics and Market Opportunities for Controlled Release Products*, O'Neill, W. P., Ed., Chemical Marketing and Economics Division, American Chemical Society, New York, 1976, 186.

Chapter 3

POLYMERS CONTAINING PENDENT PESTICIDE SUBSTITUENTS

Frank W. Harris

TABLE OF CONTENTS

I. METHODOLOGY

One approach to obtaining controlled release pesticide formulations that contain a high percentage of pesticide has been the synthesis of polymers that contain pesticides as pendent substituents.[1] For example, polymers have been prepared that consist of over 85% 2,4-dichlorophenoxyacetic acid (2,4-D) incorporated in pendent side chains.[2-4] Pesticides are released from these systems by chemically or biologically induced cleavages of the pesticide-polymer chemical bonds.

Two different synthetic routes have been employed in the preparation of polymers that contain pesticides as pendent substituents. In the first, the pesticide is converted to a polymerizable derivative that is subsequently polymerized to afford the macromolecular combination. Pesticides have also been chemically bound to preformed synthetic and naturally occurring polymers by allowing them, or one of their derivatives, to react with the polymers' functional groups.

A. Polymerization of Pesticide Derivatives

The facility with which pesticides can be converted to polymerizable derivatives depends to a large extent on their functionality. Pesticides that contain very reactive functional groups, such as carboxyl ($-CO_2H$), hydroxyl ($-OH$), sulfhydryl ($-SH$), and amino groups ($-NH_2$), can be converted to a wide variety of polymerizable derivatives. Pesticides that do not possess such functionality, e.g., diquat, trifluralin, and chlorpyrifos, cannot be used in this approach. The majority of the pesticide monomers that have been synthesized and subsequently polymerized are vinyl derivatives of carboxylic acids or alcohols.

The simplest vinyl derivative of a carboxylic acid is a vinyl ester. These compounds are readily prepared in the laboratory by the mercuric acetate-sulfuric acid catalyzed reaction of the acid with vinyl acetate.[1-8] In this reaction the vinyl group in the acetate is transferred to the pesticide replacing the acidic hydrogen.

$$ \tag{1} $$

Example:

$$ \tag{2} $$

A wide variety of polymerizable derivatives can be prepared by the reaction of the pesticidal acid with alcohols containing a vinyl group.[6,7] This reaction is catalyzed by dehydrating agents, such as dicyclohexylcarbodiimide (DCC).

$$P - CO_2H + HO - R' - CR = CH_2 \longrightarrow P - C \underset{O - R' - CR = CH_2}{\overset{O}{\diagdown}}$$

(3)

Example

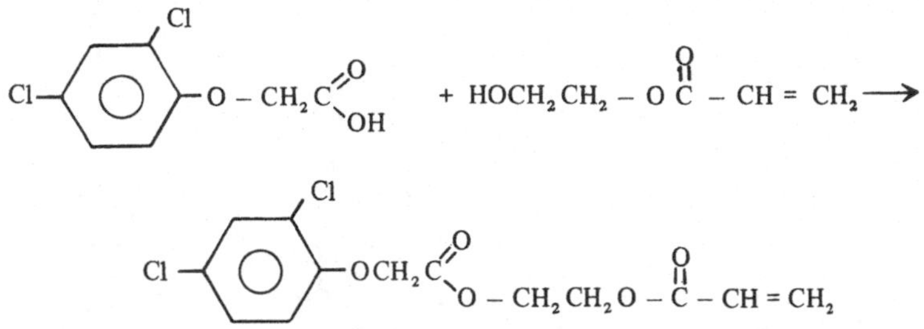

(4)

Polymerizable esters can also be prepared by a two-step process. The acid is first treated with thionyl chloride to afford the highly reactive, acid-chloride derivative. The acid chloride is then allowed to react with the alcohol in the presence of a base, such as pyridine, which neutralizes the liberated hydrogen chloride.

$$P - CO_2H \xrightarrow{\ SOCl_2\ } P - CO\,Cl$$

$$\xrightarrow[\text{Base}]{HO - R' - CR = CH_2} P - CO_2 - R' - CR = CH_2$$

(5)

Example:

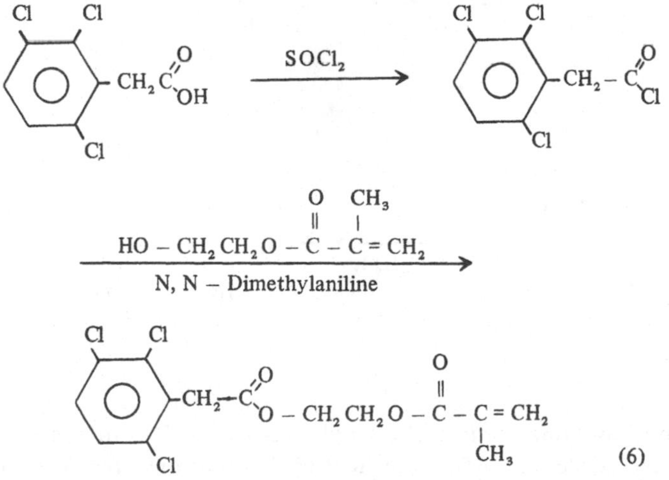

(6)

The chemistry used to prepare vinyl derivatives of pesticides containing hydroxyl groups is essentially the same as that described for carboxylic acids. The locations of the vinyl group and the pesticide are merely reversed. In this case the monomer is prepared by the reaction of a carboxylic acid, which contains the vinyl group, with the hydroxyl group of the pesticide. The ester derivatives can also be prepared by the reaction of the pesticide with the acid chloride of the carboxylic acid. Base is used to neutralize the liberated hydrogen chloride, or the reaction is carried out with the pre-formed sodium salt of the alcohol.[5]

$$P - OH + CH_2 = CR - COCl \longrightarrow CH_2 = CR - CO_2P$$

(7)

Example:

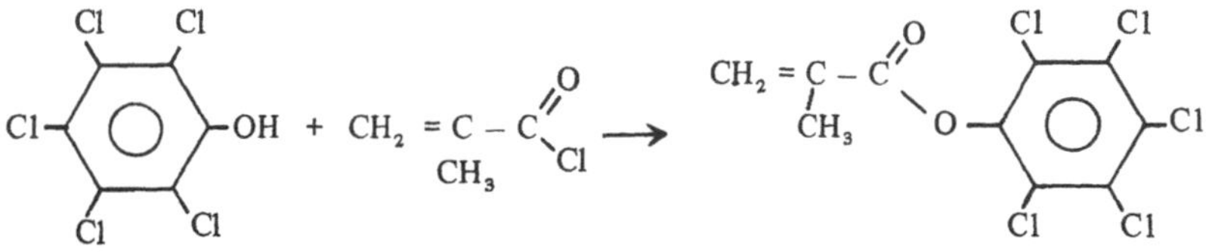

(8)

Vinyl pesticide monomers have been polymerized by bulk-, solution-, and emulsion-free radical techniques.[1-8] Bulk polymerization is the simplest method for preparing polymers. The monomer(s) is simply mixed with a small amount (<1%) of a free radical initiator, such as azobisisobutyronitrile (AIBN), and heated for several hours. The complete conversion of monomer to polymer, however, is seldom accomplished. Thus, the product is often contaminated with unreacted monomer, which can be very difficult to remove. The polymerizations are also usually exothermic, which can lead to problems with heat control. In fact, bulk polymerizations of vinyl monomers are seldom carried out with large amounts of material due to the danger of explosions.

(9)

Example:

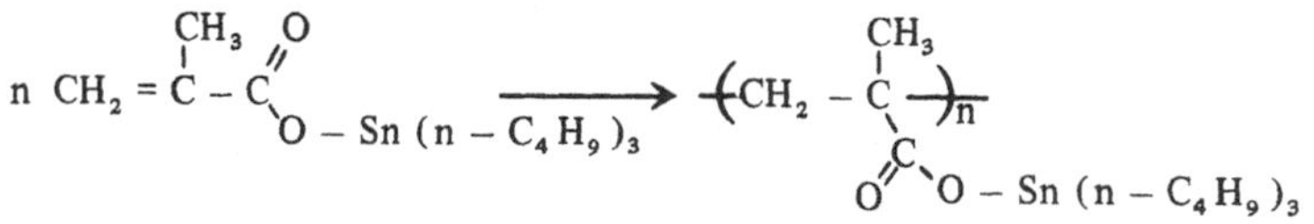

(10)

In a typical bulk polymerization, the vinyl ester of 2,4-D is mixed with 0.1% AIBN and slowly heated to 70°C.[4] After the heating is continued for 3 hr, the mixture is allowed to cool. The residue is dissolved in chloroform and then precipitated in hexane. The polymer is extracted with ethanol overnight and then dried under vacuum for 24 hr.

In solution methods the polymerization is carried out in a suitable solvent. The solvent, which must be carefully selected so as not to interfere with the polymerization, is used to take up the heat of polymerization by a rise in temperature or by vaporization. This method, however, normally affords lower molecular weight polymers. The complete removal of residual solvent from the polymer is also very difficult.

In a typical solution copolymerization, 2-acryloyloxyethyl 2,4-dichlorophenoxyacetate (95%), methacrylic acid (5%), and 0.05% AIBN are dissolved in 2-butanone and slowly heated under nitrogen to 75°C.[19] After the heating is continued for 3 hr, the mixture is cooled and added to hexane. The copolymer that precipitates is collected by filtration and dried under vacuum for 48 hr.

In emulsion polymerization, the monomer is dispersed in water by an emulsifying agent such as a detergent. Initiator radicals, usually of the redox type, are generated in the aqueous phase, and these diffuse into soap micelles that are swollen with monomer molecules. As monomer is used up in the polymerization reaction, the monomer that is dispersed in the aqueous phase migrates into the micelle. This method usually affords extremely high molecular weight materials. Emulsion polymerization also has the advantage of preparing polymers in a form suitable for direct use in coating formulations.

Vinyl 2,4-dichlorophenoxyacetate has been emulsion-polymerized by the following procedure.[8] A surfactant mixture of sodium alkylaryl polyethersulfonate (Rohm and Haas Triton® ×-200, 1.8 g) and sodium alkylaryl polyether sulfate (Rohm and Haas Triton® ×-301, 2.8 g), was added to 80 ml of distilled water. Vinyl 2,4-dichlorophenoxyacetate (20.2 g), potassium persulfate (0.2 g), and 0.7 ml of a fresh solution of ferrous sulfate heptahydrate (0.3 g in 200 ml of distilled water) were then added, and the resulting mixture was stirred at 50°C. After nitrogen was passed through the generated emulsion for 15 min, 0.2 g of sodium metabisulfite was added. The emulsion was stirred at 65°C under a nitrogen atmosphere for 1 hr, cooled, and poured into an ice-cold aqueous solution of sodium chloride contained in a Waring blender. The precipitate that formed upon stirring was collected by filtration and resuspended in the blender with fresh water. This procedure was repeated several times with fresh water and then twice with ether to afford 26.5 g of a white granular solid.

B. Reaction of Pesticides with Preformed Polymers

Another approach to polymers containing pendent pesticide substituents has been the modification of preformed synthetic or naturally occurring polymers. The method involves the reaction of a pesticide or pesticide derivative with a polymer containing appropriate functional groups. A major problem encountered in this approach is the determination of reaction conditions that will result in a high degree of pesticide substitution. As in the polymerization of pesticide monomers, the majority of the reactions that have been carried out involve carboxyl or hydroxyl groups.

Pesticidal acids can be converted to acid chlorides that will react with polymers containing pendent hydroxyl or amino groups. Both synthetic and naturally occurring polymers have been acylated in this manner.[5,8-11]

(11)

Example:

$$(12)$$

Conversely, pesticides containing hydroxyl groups will readily react with polymers containing pendent acid chlorides. Several synthetic polymers have been esterified by this technique.[12]

$$(13)$$

Example:

$$(14)$$

Polymers containing pendent hydroxyl groups have also been treated with pesticidal aldehydes. This reaction affords an acetal linkage that is susceptible to acid-catalyzed hydrolysis.[13]

$$(15)$$

Example

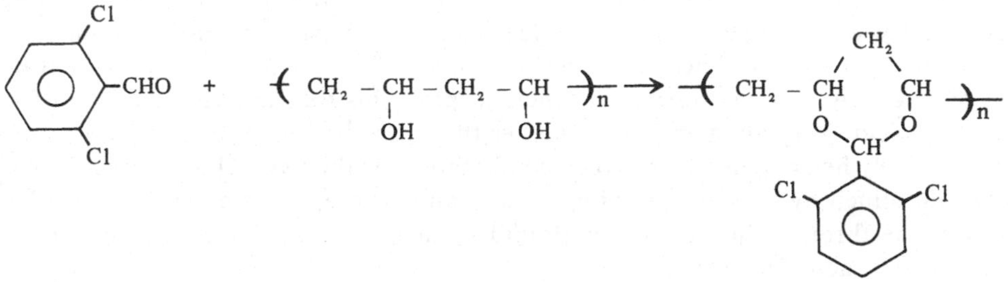

(16)

If the preformed polymer and the pesticide do not contain functionality that permits their direct reaction, they can be modified in order to affect attachment. For example, a pesticidal amine can be attached to a polymer containing pendent hydroxyl groups by first treating the pesticide with a diisocyanate.[14] The isocyanate adduct is then allowed to react with the preformed polymer.

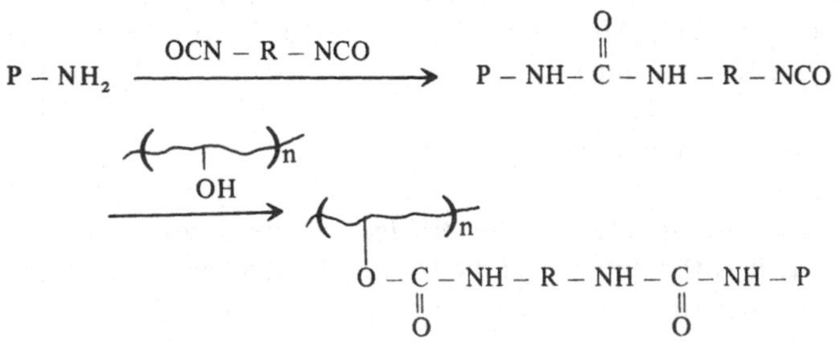

(17)

Example

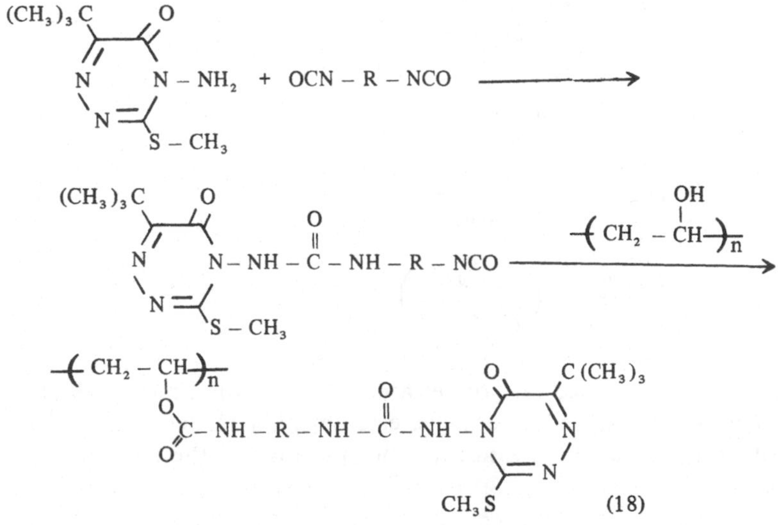

(18)

II. THEORY

As stated earlier, pendent pesticidal moieties are released from polymer backbones by chemically or biologically induced cleavages of the pesticide-polymer chemical bonds. The most common cleavage reaction employed is hydrolysis induced by water in the surrounding environment. The kinetic expressions that describe the rate of release of pesticide vary depending on whether the hydrolysis reaction is purely homogeneous, purely heterogeneous, or some combination of the two. The observed release rate is determined by both the reaction kinetics and by the diffusion rate of the liberated pesticide through the polymer.[15] Boundary layer effects can also play a role in determining the actual rate of release.[16]

If the reaction is heterogeneous, the release rate will depend on both the geometry and size of the system. Very small particles have a high surface-to-volume ratio. Hence, a heterogeneous reaction would occur faster, per unit material, than if the system consisted of larger particles.

The simplest situation mathematically is the case of a water-soluble polymer undergoing homogeneous hydrolysis in the absence of boundary layer effects. If the liberated pesticide rapidly diffuses through the polymer, the system follows conventional first-order kinetics. Diffusion times are considered rapid if they are very short compared to reaction times. In this case the release rate is reaction-rate limited.

Assuming rapid diffusion of the liberated pesticide through the polymer and ignoring boundary layer effects, Allan and Neogi derived the following rate expression for a heterogeneous reaction occurring on the surface of insoluble spherical particles:[5,17]

$$dM/dt = -\rho n \, 4\pi \, r^2 \, dr/dt = nk4\pi r^2 C_0 \tag{19}$$

where dM/dt is the rate of release of pesticide, n is the number of particles of average radius r at time t, k is the rate constant for the hydrolysis reaction, ρ is the density, and C_o is the concentration of pesticide-polymer linkages. If the hydrolyzed polymer is water-soluble, C_o is considered constant because as one pesticide molecule escapes from the surface, the water finds another pesticide-polymer linkage behind. The following equations also apply:

$$-\rho \, dr/dt = kC_0 \tag{20}$$

$$r_0 - r = \frac{k \, C_0 \, t}{\rho} \tag{21}$$

$$R_c = kn4\pi r_0^2 C_0 = kn4\pi \left(r_0 - \frac{k}{\rho} C_0 t_c\right)^2 C_0 \tag{22}$$

$$r_0 - k/p \, C_0 t_c = \left(\frac{R_c}{kn4\pi C_0}\right)^{1/2} \tag{23}$$

$$t_c = \frac{\rho r_0}{kC_0} - \left(\frac{R_c \rho^2}{nk^3 \, 4\pi C_0^3}\right)^{1/2} \tag{24}$$

where R_c is equal to the rate of release at time t_c, the time at which the activity of a particular system can no longer be detected by bioassay techniques, i.e., the persistence time of pesticidal activity. Since n is proportional to the amount of combination W, one can write

$$t_c = K_1 - K_2 W^{-1/2} \tag{25}$$

$$K_1 = \frac{\rho r_0}{k C_0} \text{ and } K_2 = \left(\frac{R_c \rho^3 r_0{}^3}{3 K^3 C_0{}^3} \right)^{1/2} \tag{26}$$

The validity of this equation was demonstrated by plotting the persistence time (t_c) of a herbicide/polymer chemical system, which was synthesized by the partial acylation of poly(vinyl alcohol) with 2-methyl-4-chlorophenoxyacetyl chloride, against the inverse of the square root of the amount of formulation used ($W^{-1/2}$).

Neogi and Allan also derived the following expressions for the rate of cleavage when the hydrolyzed polymer is water-insoluble, e.g., cellulose or lignin:[5,10,17]

$$-dc/dt = k_2 C \tag{27}$$

$$-dc/C = k_2 dt \tag{28}$$

$$\ln C_0/C = k_2 t \tag{29}$$

where C is the concentration of pesticide per unit weight at time t, and k_2 is the degradation rate constant. In this case

$$R_c = k_2 W C_0 e^{-k_2 t_c} \tag{30}$$

and

$$t_c = \frac{2.3}{k_2} \log W - \frac{2.3}{k_2} \log R_c/k_2 C_0 \tag{31}$$

The validity of this equation was demonstrated by plotting the persistence time (t_c) of 2,4-D/cellulose and 2,4-D/kraft lignin chemical combinations against the log of the amount used (log W).

Baker and Lonsdale have considered a theoretical case where heterogeneous or homogeneous reactions occur while diffusion of the liberated pesticide is prohibited.[15] In their example, a homogeneous reaction, such as hydrolysis, is occurring throughout the bulk with a rate constant k_b. Simultaneously, a reaction is occurring in the surface layer of thickness δ with a rate constant k_s. As a result of the surface reaction, the surface erodes, for example, by dissolution.

The two rate equations are written:

$$k_s A = -dA/dt \tag{32}$$

for the surface reaction, and

$$k_b A = -dA/dt \tag{33}$$

for the bulk reaction, where A is the concentration of labile groups present at any time. After time t, the polymer in the surface layer will have reacted for a time t_b at the bulk rate and t_s at the surface rate. Hence, in the surface layer

$$t = t_b + t_s \tag{34}$$

From equations (32) and (33) it follows that:

$$t_b = \frac{1}{k_b} \int_{A^o}^{A'} dA/A = \frac{1}{k_b} \ln (A^o/A') \qquad (35)$$

and

$$t_s = \frac{1}{k_s} \int_{A'}^{A''} dA/A = \frac{1}{k_s} \ln (A'/A'') \qquad (36)$$

where A^o is the initial concentration of reactable bonds, A' is the concentration of bonds at the interface between the bulk and the surface layer, and A'' is the concentration of bonds at which the polymer erodes. This could be the water-solubility limit, for example. Some unknowns can be eliminated by using the approximate relationship

$$dx/dt = \delta/t_s \qquad (37)$$

Combining Equations 34 to 37 leads to the expression for the erosion rate, dx/dt:

$$dx/dt = \frac{(k_s - k_b) \delta}{\ln (A^o/A'') - tk_b} \qquad (38)$$

Thus, if the two rate constants are known along with the initial concentration of bonds and the concentration at which the polymer becomes soluble, the erosion rate can be calculated. Assuming the pesticide loading is known, the release rate can then be determined.

Two limiting cases of Equation 38 exist. When $k_b = 0$, i.e., for complete surface erosion, the erosion rate is constant and given by

$$dx/dt = \frac{\delta k_s}{\ln (A^o/A'')} \qquad (39)$$

When $k_s = k_b$, i.e., when there is no preferential surface reaction, there is no erosion of the device at all until the bulk concentration reaches A'' at which point erosion occurs at an infinite rate. The lifetime in this case can be obtained from Equation 38:

$$t_\infty = \frac{1}{k_b} \ln (A^o/A'') \qquad (40)$$

The lifetime is independent of the thickness or shape of the system in this limit.

A plot of the fractional erosion, x/ℓ, for a slab of thickness ℓ against reduced time, $t/t\infty$, for various ratios of the rate constants k_s/k_b is presented in Figure 1. The two limiting cases are apparent: for no bulk reaction, the erosion rate is linear with time, i.e., zero-order kinetics are followed. With no surface reaction, the system suddenly disintegrates at $t\infty$. Although there are no experimental data for systems of this type, the analysis has utility in that it shows possible patterns of release.

There are no mathematic treatments in the literature of situations where diffusion times and reaction times are comparable. Work in this area is needed because, as Baker and Lonsdale have pointed out, not only is diffusion likely to be important in real

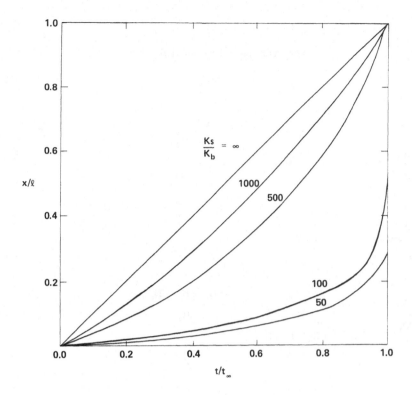

FIGURE 1. Fractional erosion vs. reduced time for various ratios of the surface and bulk reaction rate constants.

systems, but concentration- or time-dependent diffusion coefficients can be anticipated.

In all systems the value of the hydrolysis rate constant depends on the strength and chemical nature of the pesticide-polymer chemical bonds, the structure of the polymer, and the environmental conditions. For example, an anhydride linkage is more susceptible to hydrolysis than an ester or amide linkage. The rate of hydrolysis of a linkage is also dependent on the groups surrounding it. Thus, hydrophilic neighboring groups enhance hydrolysis while hydrophobic groups can slow and even prevent reaction. In fact, it appears that some hydrophilic groups must be present in order for hydrolysis to occur under mild conditions. There is also evidence that in cases of enzyme-catalyzed hydrolysis the hydrophilicity of the system is more important than the bonding energy in determining the overall rate of release.[8] An uncrosslinked polymer is much more susceptible to hydrolysis than a highly crosslinked one, and a crystalline polymer is less susceptible to hydrolytic attack than an amorphous polymer. It does appear that the release rate is relatively independent of the polymer's molecular weight. Of course, the temperature and the pH of the surrounding medium will significantly affect the rate of hydrolysis with high temperature and both high and low pH favoring the reaction.

III. APPLICATIONS

A. Controlled Release Herbicides and Fungicides

Although no commercial products have been offered based on polymers containing pendent herbicide and fungicide substituents, several promising products are in various

TABLE 1

Polymerization of Pesticide Derivatives

Pesticide derivative	Polymer structure	Biologically active
2-chloro-methylphenyl ring with $-OCH_2-CO_2CH=CH_2$, CH_3 substituent	$-(CH_2-CH)-$ with O, $C-P$ (ester, $C=O$)	No
2-chloro-methylphenyl ring with $-O-CH_2CO_2CH=CH_2$, CH_3 substituent	$-(CH_2-CH)-(CH_2-CH)-$ with O/$C-P$ and CO_2H	Yes
2-chloro-methylphenyl ring with $-O-CH_2CO_2CH_2CH-CH_2$ (epoxide O), CH_3 substituent	$-(O-CH-CH_2)-$ with CH_2, O, $C-P$ ($C=O$)	Yes
tetrachloro-methyl phenyl ring with $-O-C(=O)-C(CH_3)=CH_2$	$-(CH_2-C)-$ with CH_3, C, O $O-P$	No
tetrachloropyridyl (N) ring with $-O-C(=O)-C(CH_3)=CH_2$	$-(CH_2-C)-$ with CH_3, C, O OP	No
tetrachloropyridyl (N) ring with $-O-C(=O)-C(CH_3)=CH_2$	$-(CH_2-C)-(CH_2-CH)-$ with CH_3, C, O OP and CO_2H	Yes

stages of development. The following is a brief review of the published research in this area of controlled release technology.

1. Polymerization of Herbicide and Fungicide Derivatives

A series of monomers containing terrestrial herbicides has been synthesized, polymerized, and the biological activity of the resulting polymers determined (Table 1).[5,17] The results of this study indicated that in order for these polymeric herbicides to display biological activity they must contain some hydrophilic groups to enhance the degradation rate. Incomplete data were given for the periods of control afforded by these

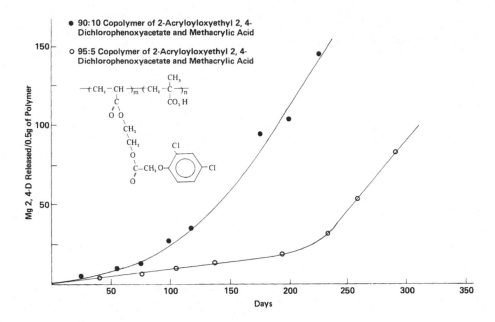

FIGURE 2. Hydrolysis of methacrylic acid copolymers.

formulations. However, in the case of the acrylic acid-2,3,5-trichloro-4-pyridylmethacrylate copolymer, increasing the percentage of acrylic acid resulted in shorter periods of control. It was postulated that the pendent carboxyl groups serve as catalysts for hydrolysis of the neighboring ester linkages.

The herbicides 2,4-D and 2-(2,4,5-trichlorophenoxy)propionic acid (Silvex®) have been incorporated in the alcohol residues of a series of acrylic esters.[4,6,7] In these monomers the distance between the herbicide and the vinyl group and, hence, the resulting polymer backbone was varied by varying the length of the unit connecting the herbicide to the acrylic acid. It was postulated that increasing the length of the pendent side chain would enhance the hydrolysis of the herbicide-polymer bond since the ester linkage would be removed from the hydrophobic backbone and less sterically hindered.

The monomers, i.e., 2-acryloyloxyethyl 2,4-dichlorophenoxyacetate, 2-acryloyloxyethyl 2-(2,4,5-trichlorophenoxy)propionate, 2-methacryloyloxyethyl 2,4-dichlorophenoxyacetate, 2-methacryloyloxyethyl 2-(2,4,5-trichlorophenoxy)propionate, 4-acryloyloxybutyl 2,4-dichlorophenoxyacetate, 4-acryloyloxybutyl 2,4-dichlorophenoxyacetate, and 4-methacryloyloxybutyl 2-(2,4,5-trichlorophenoxy)propionate, were polymerized by bulk- and solution-free radical techniques to yield the corresponding polymers. Hydrolysis studies, however, showed that none of the homopolymers will undergo hydrolysis under mildly alkaline conditions at 30°C.

In a more successful approach to obtaining hydrolyzable systems, the 2-acryloyloxyethyl and 2-methacryloyloxyethyl esters of 2,4-D and Silvex® were copolymerized with varying amounts of hydrophilic monomers, such as methacrylic acid and trimethylamine methacrylimide.[18,19] The 2-acryloyoxyethyl ester copolymers slowly underwent hydrolysis under slightly alkaline conditions (pH = 8) at 30°C. Increasing the percentage of the hydrophilic comonomer in the system resulted in an increase in the rate of herbicide release (Figure 2). As can be seen in the figure, the rates of hydrolyses of the copolymers increased with time. This was attributed to intramolecular interactions of the remaining ester groups with carboxyl groups generated on the chain during the hydrolyses.

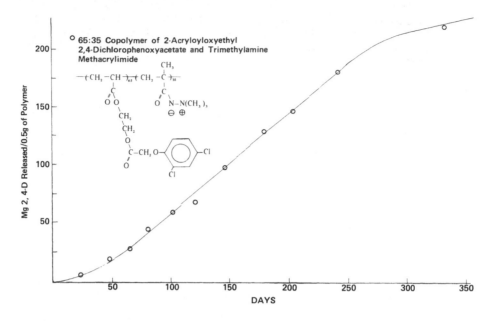

FIGURE 3.   Hydrolysis of trimethylamine methacrylimide copolymer.

The release rate profile obtained for a copolymer of 2-acryloyloxyethyl 2,4-dichlorophenoxyacetate containing 35 mol% trimethylamine methacrylimide offers particular promise (Figure 3). As can be seen from Figure 3, the rate of hydrolysis increased during the first few days of the study and then remained relatively constant. This behavior, in effect, provides a zero-order rate of release, which is ideally suited for controlled release applications.

Surprisingly, the 2-methacryloyloxyethyl ester copolymers did not undergo hydrolysis under the mildly alkaline conditions. This decrease in reactivity may be due to steric hindrance effects.

Three organic fungicides, pentachlorophenol, 2-(4'-thiazoyl)benzimidazole, and 8-hydroxyquinoline, have been converted into acrylate monomers and then copolymerized in bulk with either ethyl acrylate or vinyl acetate.[20] The copolymers' biological activities are currently being evaluated in a microbiological accelerated growth testing program. Several of the copolymers have resisted mildew growth from agar media where control samples of poly(ethyl acrylate) and poly(vinyl acetate) failed.

A hard friable alkyd resin has been produced by the addition of 2,4-dichlorophenoxyacetic acid to a glycerol and phthalic anhydride or humic acid reaction mixture. The longer effective lifetime of the herbicide was attributed to its protection from rapid bacterial decomposition.[21]

2. Reaction of Herbicides and Fungicides with Preformed Polymers

A synthetic polymer containing a terrestrial herbicide has been prepared by the partial acylation of poly(vinyl alcohol) with 2-methyl-4-chlorophenoxyacetyl chloride.[5] The duration of herbicide release from the polymer was determined through observation of the inhibition of seed germination in lettuces which were sown daily. Control was achieved for periods of up to 100 days.

Poly(vinyl alcohol) has also been treated with 2,6-dichlorobenzaldehyde (DCBA) in the presence of acid to afford the corresponding poly(vinyl acetal) copolymer.[13] The maximum degree of acetalization obtained was 68%. The copolymers, however, did not undergo hydrolysis in water with a pH = 1 at 25°C.

A series of laboratory and commercial polymers of poly(vinyl alcohol), which contained varying residual amounts of unhydrolyzed vinyl acetate, have been treated with isocyanate adducts of metribuzin to afford the corresponding copolymers.[14] The metribuzin adducts were obtained from the reactions of the herbicide with diisocyanates. Both linear and crosslinked copolymers were prepared with varying degrees of metribuzin substitution. Results of release rate studies conducted in distilled water showed that the linear polymers release the herbicide much faster than crosslinked systems. Soil thin-layer chromatographic experiments indicated that the soil mobility of metribuzin is greatly decreased by such chemical attachment, while a bioassay study showed that its residual phytotoxicity is significantly increased.

Poly(ethyleneimine) samples with low and medium molecular weights have been acylated with 2,4-D.[8] Polymers were prepared with varying degrees of substitution. The release rates of the amide derivatives, which were determined by bioassay in soil, show a wide range with no correlation with molecular weight. The degree of substitution of 2,4-D did influence the release rate, with the less substituted polymers having faster release rates. The amide-bonded herbicides were released faster than were herbicides from corresponding ester-bonded formulations. These results suggest that the system's hydrophilicity, which affects the hydrolytic enzyme's accessibility to the linking bond, is more important in determining the overall release rate than the bonding energy.

Herbicides have also been covalently bonded by ester or amide linkages to a divinylbenzene-styrene copolymer.[22] Esterification was carried out by the reaction of the triethylammonium salt of 2,4-dichlorophenoxyacetic acid with a chloromethylated resin. Amidation was accomplished by the acylation of an aminomethylated resin with 2,4-dichlorophenoxyacetic acid. Dicyclohexylcarbodiimide was used to catalyze the reaction. Herbicide reactivation was produced enzymatically using lipase, acetylcholinesterase, and trypsin.

Mildew-resistant resins have been prepared by treating alkyd resins with tri-n-butyltin oxide.[23,24] These resins were subsequently incorporated in experimental paint formulations. The paints showed good mildew resistance in chamber tests.

Several naturally occurring polymers have also been used in the preparation of controlled release herbicide-polymer formulations.[5,17,25,26] For example, the cellulose in Douglas fir bark and Kraft lignin has been acylated with 2,4-dichlorophenoxyacetyl chloride. The duration of herbicide release from these formulations was determined by observation of the inhibition of seed germination in lettuces which were sown daily. A 20% 2,4-dichlorophenoxyacetic acid/Douglas fir bark chemical combination afforded control for approximately 100 days, while a comparable treatment with a 39% 2,4-dichlorophenoxyacetic acid/Kraft lignin formulation prevented seed germination for 170 days. Similarly, long term suppression of competitive vegetation in reforestation efforts has been accomplished by employing bark which has been treated with 2,4-dichlorophenoxybutyric acid (2,4-DB).[27-29]

Pentachlorophenol has also been chemically bonded to the cellulose fiber surfaces of softwood Kraft pulp by S-triazine bridges, which were formed by a cyanuric chloride treatment in an aqueous medium. Such fibers were completely protected from fungal atack by *Lenzites trabea* for 49 days.[30]

Unmodified, cyanoethylated, and crosslinked corn starches have been treated with 2,4-dichlorophenoxyacetyl chloride in pyridine to produce the corresponding 2,4-D esters.[31] Hydrolysis studies at pH 6 and 8 at ambient temperature showed that these insoluble compositions quickly liberate 2,4-D and soluble 2,4-D esters by ester and glucosidic bond cleavages, respectively.

B. Controlled Release Antifoulants

Presently two companies market antifouling paints based on polymers containing pendent organotin substituents. Nippon Oils and Fats Co., Ltd. sells under the Takata® label antifouling paints formulated with a copolymer of tri-*n*-butyltin methacrylate and other acrylic monomers.[32] International Red Hand® also produces an organotin polyacrylic based antifouling paint, SPC-40.[33] Both systems are proposed to work by the release of the organotin antifoulant at the surface, followed by dissolution of the surface layer to expose a new bioactive layer of polymer. M & T Chemicals, Inc. also has available developmental quantities of a series of organotin methacrylate polymers. Marine formulations based on these polymers are currently being evaluated in ship trials. In addition to these commercial products, several other promising antifouling formulations are in various stages of development. The following is a brief review of the published research with antifouling polymers.

1. Polymerization of Antifoulant Derivatives

Pentachlorophenyl methacrylate has been prepared by the reaction of pentachlorophenol with methacryloyl chloride.[11] This monomer was then converted to the corresponding polymer. Ocean immersion tests showed that the polymer provides long-term fouling protection.

A series of antifouling polymers containing pendent organometallic substituents has been prepared by the copolymerization of organometallic monomers, such as tributyltin methacrylate, with unsaturate comonomers or resins.[34-39] The organometallic groups attached to the polymer backbones included tributyltin, tripropyltin, trimethyltin, triphenyltin, and tribenzyltin moieties. The resins were found to be most effective against fouling organims when the organic radicals of the organometallic moieties were propyl or butyl groups. The antifouling activity of the polymers were broadened by incorporating two or more different organometallic groups along their backbones. For example, poly(tri-*n*-butyltin methacrylate-co-tri-*n*-propyltin methacrylate-co-methyl methacrylate) has given excellent antifouling performance for several years in ocean immersion tests.

The relative rates of release of total tin from several commercial organotin salt-based antifouling paints and poly(tri-*n*-butyltin methacrylate-co-tri-*n*-propyltin methacrylate-co-methyl methacrylate) have been determined in an accelerated aging device.[38] The organotin paints released tin two to three times faster than the polymer. Studies to determine the organotin polymer's rate of release under use conditions, the mechanism of release, and the identity of the released species are currently in progress.

Poly(tri-*n*-butyltin methacrylate-co-methyl methacrylate) and poly(tri-*n*-butyltin methacrylate-co-tri-*n*-propyltin methacrylate-co-methyl methacrylate) have been formulated into antifouling paints.[40] Although both polymers exhibit fair film forming characteristics, they produced hard paint films only upon addition of special crosslinking agents.

Organometallic polyacrylates were originally developed as antislime, transparent coatings for optical surfaces.[41] Resins such as poly(tri-*n*-butyltin methacrylate-co-methyl methacrylate) combine the transparency and film-forming properties of acrylics with the antifouling activity of organometallic compounds. Water and oil repellency have been imparted to these coatings by copolymerizing the organometallic acrylates with fluoroalkyl acrylates. Evaluation of copolymers, such as poly(1H,1H-pentadecafluorooctyl methacrylate-co-tributyltin methacrylate) as no-wet, antifoulant, transparent plastics are now in progress.

Hydrophilic comonomers, such as 2-hydroxyethyl acrylate, have also been copolymerized with organotin acrylate monomers.[39] The hydrophilic groups were incorporated in the polymeric systems in an attempt to produce drag-reducing coatings with antifouling properties. These systems are currently being evaluated.

2. Reaction of Antifoulants with Preformed Polymers

A series of antifouling polymers has been prepared by the reactions of phenarsazine chloride, pentachlorophenol, 2,4,6-trichlorophenol, tri-n-butyltin hydroxide, 1-hydroxyimidazoles and 1-hydroxymidazole-3-oxides with polyacryloyl chloride and polymethacryloyl chloride.[11,12] The polymers antifouling properties were evaluated by the mortality of pond snails. The polymers that contained several different toxicants were more effective than physical mixtures of polymers which contained monotoxicants. It was postulated that the improved antifouling properties were due to synergistic interactions of the toxicants of the polymer backbone.

Pentachlorophenol, phenylazo-2-naphthol, p-hydroxyphenylazo-2-naphthol, phenarsazine chloride and triethyltin hydroxide have also been allowed to react with a styrene-maleic anhydride copolymer.[11] Ocean immersion tests showed that the polymerized toxicants provide long-term fouling protection.

Polyacylic acid, polymethacrylic acid, carboxymethyl cellulose, the sodium salt of carboxymethyl cellulose, and a methyl vinyl ether-maleic acid copolymer have been treated with trialkyltin oxides and triaryltin hydroxides to afford a series organotin-containing polymers.[38] Poly-1,2-butadiene has also been treated with tributyltin hydride in the presence of a free radical catalyst to give the corresponding substituted polymer. In these syntheses, organometallic polymers of different physical characteristics were obtained by varying the backbone and the molecular weight of the polymer. The toxic properties of the prepared polymers were altered by incorporating different groups onto the polymer backbone and by varying the molar ratio of the organometallic substituent. In field-exposure trials conducted at Pearl Harbor, the acrylic- and poly(methyl vinyl ether-co-maleic acid)-based systems that contained a variety of organometallic substituents gave the best antifouling performance.

Tri-n-butyltin esters of poly(methyl vinyl ether-co-maleic acid) and crosslinked polymethacrylic acid have been incorporated in rubber-based antifouling paint formulations.[40] The vinyl organotin ester was only marginally compatible with the rubber resins and produced soft, tacky films. After particle size reduction, the crosslinked acrylic system was successfully substituted for conventional pigments in various paint formulations.

Polyester and epoxy resins containing organotin substituents are currently being developed for use in the fabrication of inherently antifouling glass-reinforced composites.[42,43] These materials could be used in the construction of many marine structures, such as buoys and glass reinforced-plastic seawater piping systems. The organotin moieties are incorporated in the polyester and epoxy systems by use of organotin-containing curing agents. For polyesters, tributyltin methacrylate is used in combination with styrene to crosslink preformed, unsaturated alkyd resins. For epoxies, the organotin curing agent is produced by partially esterifying pendent acid or anhydride groups on a vinylic polymer. Preformed bisphenol A, novolac, and cycloaliphatic epoxy resins are then cured with the organotin curing agent. These organometallic polyester and epoxy resins have demonstrated 100% antifouling performance after 17 and 12 months of ocean exposure, respectively.

Glass-reinforced laminates have been fabricated using the organotin polyester and epoxide resins.[42] The polyester laminates, which were prepared by vacuum bag techniques, showed minimal reduction in structural properties when compared to conventional polyester laminates. Two types of epoxy laminates were prepared. A shell laminate, which was prepared by laying up the bottom plies with a commercial epoxy system and the remaining top plies with the organotin epoxy system, showed less than a 10% decrease in structural properties. The antifouling properties of the glass-reinforced laminates are currently being evaluated.

Naturally occurring polymers have also been used in the preparation of antifouling formulations.[44] Tributyltin oxide, 2,4-dichlorophenoxyacetic acid and pentachlorophenol were allowed to react with waste polymers such as ligno-sulfonates, Kraft lignin, bark fiber, crab shell, chitin, and alginic acid. The resulting polymeric formulations exhibited better antifouling properties than the original monotoxicants. In fact, a chronic intoxication of sea weeds was induced through their long-term exposure to the low concentrations of the slowly released toxicants.

IV. ACKNOWLEDGMENT

I would like to acknowledge the continued support of the author's research in this area by the U.S. Army Engineer Waterways Experiment Station, Vicksburg, Miss. I would also like to thank Ms. Jean A. Montemarano, David W. Taylor Naval Ship Research & Development Center, Annapolis, Md., for her invaluable assistance in compiling the antifouling section of this chapter.

REFERENCES

1. **Feld, W. A., Post, L. K., and Harris, F. W.**, Controlled release from polymers containing pesticides as pendent substituents, in *Proc. 1975 Int. Controlled Release Pesticide Symp.*, Harris, F. W., Ed., Wright State University, Dayton, Ohio, 1975, 113.
2. **Dovlatyan, V. V. and Kostanyan, D. A.**, Polyvinyl esters of aryloxyacetic acids, *Izv. Akad. Nauk Arm. SSR, Khim. Nauki*, 18(3), 325, 1965; *Chem. Abst.*, 64, 623, 1966.
3. **Harris, F. W. and Post, L. K.**, Synthesis and polymerization of the vinyl and acryloyloxyethyl esters of 2,4-dichlorophenoxyacetic acid and 2-(2,4,5-trichlorophenoxy) propionic acid, *J. Polym. Sci., Part A-1*, 13, 225, 1975.
4. **Harris, F. W. and Post, L. K.**, Synthesis of polymers containing pendent herbicide substituents, *Am. Chem. Soc., Polym. Prepr.*, 16(2), 622, 1975.
5. **Neogi, A. N.**, Polymer Selection for Controlled Release Pesticides, Ph. D. thesis, University of Washington, Seattle, 1970.
6. **Harris, F. W. and Post, L. K.**, Synthesis of polymers containing aquatic herbicides as pendent substituents, in *Proc. 1974 Int. Controlled Release Pesticide Symp.*, Cardarelli, N. F., Ed., University of Akron, Akron, Ohio, 1974, 17.
7. **Harris, F. W., Feld, W. A., and Bowen, B.**, Polymers containing pendent herbicide substituents, in *Proc. 1975 Int. Controlled Release Pesticide Symp.*, Harris, F. W., Ed., Wright State University, Dayton, Ohio, 1975, 334.
8. **Wilkins, R. M.**, Design and evaluation of polymer combinations for the controlled release of herbicides, in *Proc. 1976 Int. Controlled Release Pesticide Symp.*, Cardarelli, N. F., Ed., University of Akron, Ohio, 1976, 7.
9. **Allan, G. G., Chopra, C. S., Neogi, A. N., and Wilkins, R. M.**, Design and synthesis of controlled release pesticide-polymer combinations, *Nature (London)*, 234, 349, 1971.
10. **Allan, G. G., Beer, J. W., and Cousin, M. J.**, Controlled release of herbicides from biodegradable substrates, in *Controlled Release Pesticides*, ACS Symp. Ser. No. 53, Scher, H. B., Ed., American Chemical Society, Washington, D.C., 1977, 94.
11. **Akagane, K. and Matsuura, K.**, Long-term protection. I. Effects of toxic polymers on long-term protection, *Shikizai Kyokaishi*, 45(2), 69, 1972; *Chem. Abst.*, 77, 128180r, 1972.
12. **Akagane, K. and Allan, G. G.**, Long-term protection. V. Mutual interactions of active toxicants in antifouling polymers consisting of multicomponent toxicants, *Shikizai Kyokaishi*, 46(8), 437, 1973; *Chem. Abst.*, 80, 61104 k, 1974.

13. **Schacht, E. H., St. Pierre, T., Demarets, G. E., and Goethals, E. J.,** Polymeric pesticides. I. Synthesis and characterization of 2,3-dichlorobenzaldehyde-generating polymers, *Am. Chem. Soc. Polym. Prepr.,* 18(1), 590, 1977.

14. **McCormick, C. L. and Fooladi, M.,** Synthesis, characterization, and release mechanisms of polymers containing pendent herbicides, in *Controlled Release Pesticides,* ACS Symp. Ser. No. 53, Scher, H. B., Ed., American Chemical Society, Washington, D.C., 1977, 112.

15. **Baker, R. W. and Lonsdale, H. K.,** Principles of controlled release, in *Proc. 1975 Int. Controlled Release Pesticide Symp.,* Harris, F. W., Ed., Wright State University, Dayton, Ohio, 1975, 9.

16. **Roseman, T. J.,** Boundary layer considerations in the evaluation of controlled release delivery systems, in *Proc. 1977 Int. Controlled Release Pesticide Symp.,* Goulding, R. L., Ed., Oregon State University, Corvallis, 1977, 403.

17. **Neogi, A. N. and Allan. G. G.,** Controlled-release pesticides: concepts and realization, in *Advances in Experimental Medicine and Biology No. 47, Controlled Release of Biologically Active Agents,* Tanquary, A. C. and Lacey, R. E., Eds., Plenum Press, New York, 1974, 195.

18. **Harris, F. W., Aulabaugh, A. E., Case, R. D., Dykes, M. R., and Feld, W. A.,** Polymers containing pendent herbicide substituents: preliminary hydrolysis studies, in *Controlled Release Polymeric Formulations,* ACS Symp. Ser. No. 33, Paul, D. R. and Harris, F. W., Eds., American Chemical Society, Washington, D.C., 1976, 222.

19. **Harris, F. W., Dykes, M. R., Baker, J. A., and Aulabaugh, A. E.,** Polymers containing pendent herbicide substituents: hydrolysis studies II, in *Controlled Release Pesticides,* ACS Symp. Ser. No. 53, American Chemical Society, Washington, D.C., 1977, 102.

20. **Stahl, C. A. and Pittman, C. U.,** Chemically anchoring mildewcides to model paint binders, *Am. Chem. Soc., Div. Org. Coat. Plast. Chem. Pap.,* 37(1), 355, 1977.

21. **Baltazzi, E.,** Herbicidally active alkyd resins, U.S. Patent 3,343,941; *Chem. Abstr.,* 68, 50730n, 1968.

22. **Jakubke, H. D. and Busch, E.,** Synthesis of low-molecular weight active substances covalently bonded to polymeric supports and investigations of enzymic reactivation, *Z. Chem.,* 13(3), 105, 1973; *Chem. Abstr.,* 79, 1198 p, 1973.

23. **Steel, M. D. and Drisko, R. W.,** Fungal-resistant organotin resins, *J. Coat. Tech.,* 48, 59, 1976.

24. **Drisko, R. W., Schwab, L. K., and O'Neill, T. B.,** Controlled release of organotin biocides from mildew-resistant paints, in *Proc. 1977 Controlled Release Pesticide Symp.,* Goulding, R. A., Ed., Oregon State University, Corvallis, 1977, 181.

25. **Allan, G. G.,** Modification reactions (of lignin), in *Lignins,* Sarkanan, K. V., Ed., Interscience, New York, 1971, 511.

26. **Allan, G. G., Chopra, C. S., and Russel, R. M.,** Controlled release pesticides. III. Selective suppression of weeds and deciduous brush in the presence of conifers, *Int. Pest Control,* 14(2), 15, 1972.

27. **Allan, G. G., Friedhoff, J. F., and Powell, J. C.,** Controlled release pesticides. VIII. Growth enhancement of a juvenile conifer forest four years after application of a controlled release herbicide, *Int. Pest Control,* 17(2), 4, 1975.

28. **Allan, G. G., Cousin, M. J., McConnell, W. J., Powell, J. C., and Yahiaoui, A.,** Controlled release herbicides in forestry, in *Proc. 1976 Controlled Release Pesticide Symp.,* Cardarelli, N. F., Ed., University of Akron, Ohio, 1976, 7.

29. **Allan, G. G., Beer, J. W., and Cousin, M. J.,** Current status of controlled release herbicides in Pacific-Northwest reforestation, in *Proc. 1977 Int. Controlled Release Pesticide Symp.,* Goulding, R. L., Ed., Oregon State University, Corvallis, 1977, 19.

30. **Allan, G. G. and Halabisky, D. D.,** Fiber surface modifications. XI. Covalent attachment of fungicides to lignocellulosic fibers, *J. Appl. Chem. Biotechnol.,* 21(7), 190, 1971.

31. **Mehltretter, C. L., Roth, W. B., Weakley, F. B., McGuire, T. A., and Russell, C. R.,** Potential controlled-release herbicides from 2,4-D esters of starches. *Weed Sci.,* 22(5), 415, 1974.

32. **Antifouling Paint,** United Kingdom Patent 1,062,324, March 22, 1967.

33. **Improved Antifouling Composition,** United Kingdom Patent 1,124,297, Aug. 21, 1968.

34. **Montemarano, J. A. and Dyckman, E. J.,** Biologically active polymeric materials exhibiting controlled release mechanisms for fouling prevention, in *Proc. 1974 Int. Controlled Release Pesticide Symp.,* Cardarelli, N. F., Ed., University of Akron, Ohio, 1974, 21.

35. **Montemarano, J. A. and Dyckman, E. J.,** Performance of organometallic polymers as antifouling materials, *J. Paint Technol.,* 47(600), 59, 1975.

36. **Dyckman, E. J., Montemarano, J. A., and Fischer, E. C.,** Environmentally compatible antifouling materials: organometallic polymers, *Nav. Eng. J.,* 33, December 1973.

37. **Castelli, V. J. and Yeager, W. L.,** Organometallic polymers: development of controlled release antifoulants, in *Controlled Release Polymeric Formulations,* ACS Symp. Ser. No. 33, Paul, D. R. and Harris, F. W., Eds., American Chemical Society, Washington, D.C., 1976, 239.

38. **Dyckman, E. J., Montemarano, J. A., Anderson, D. M., and Miller, A. M.,** Nonpolluting antifouling organometallic polymers, Rep. 3581, David W. Taylor Naval Ship Research and Development Center, Annapolis, Md., November 1972.

39. **Castelli, V. J., Anderson, D. M., Mullin, C. E., and Yeager, W. L.,** Polymers for antifouling drag-reducing coating systems, Rep. MAT-76-20, David W. Taylor Naval Ship Research and Development Center, Annapolis, Md., May 1976.

40. **Yeager, W. L. and Castelli, V. J.,** Antifouling applications of various tin containing organometallic polymers, paper presented at 173rd National American Chemical Society Meeting, New Orleans, La., March 1977.

41. **Montemarano, J. A. and Dyckman, E. J.,** Antifouling organometallic structural plastics, Rep. 4159, David W. Taylor Naval Ship Research and Development Center, Annapolis, Md., August 1973.

42. **Montemarano, J. A. and Cohen, S. A.,** Antifouling glass-reinforced composite materials, Report MAT-75-33, David W. Taylor Naval Ship Research and Development Center, Annapolis, Md., January 1976.

43. **Subramanian, R. V. and Gary, B. K.,** Controlled-release antifouling coatings based upon organotin epoxide polymers, in *Proc. 1977 Controlled Release Pesticide Symp.,* Goulding, R. A., Ed., Oregon State University, Corvallis, 1977, 154.

44. **Akagane, K. and Allan, G. G.,** Antifouling polymers for controlling sea weeds, IV, *Shikizai Kyokaishi,* 46(6), 370, 1973; *Chem. Abst.,* 79, 101570X, 1973.

Chapter 4

MICROENCAPSULATION USING COACERVATION/PHASE SEPARATION TECHNIQUES

Joseph A. Bakan

TABLE OF CONTENTS

1. ESTABLISHMENT OF THREE-PHASE SYSTEM

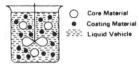

○ Core Material
● Coating Material
≡ Liquid Vehicle

2. DEPOSITION OF LIQUID-POLYMERIC COATING MATERIAL

3. SOLIDIFICATION OF COATING MATERIAL

FIGURE 1.   Schematic of processes.

I. INTRODUCTION

Microencapsulation by coacervation-phase separation is generally attributable to the NCR Corporation and the patents of Green et al.[1-9]

The microencapsulation processes can be described as techniques to reproducibly apply uniformly thin, polymeric coatings to small particles of solids, droplets of pure liquids or solutions, and dispersions. Microcapsules, the result of the processes, range in size from several tenths of a micron to a few thousand microns. The core material (syn., fill, internal phase, [IP]) is the particulate mass to be microencapsulated. The coating material has also been referred to as the shell or wall material in numerous publications.

II. GENERAL DESCRIPTION OF PROCESS

The process consists of a series of three steps carried out under continuous agitation: (1) formation of three immiscible chemical phases; (2) deposition of the coating; and (3) rigidization of the coating (Figure 1).

Step 1 — Formation of three, immiscible, chemical phases: a liquid manufacturing vehicle phase, a core material phase, and a coating material phase. To form the three phases, the core material is dispersed in a solution of the coating polymer, the solvent for the polymer being the liquid manufacturing vehicle phase. The coating material phase, an immiscible polymer in a liquid state, is formed by utilizing one of the methods of phase separation-coacervation, that is, by changing the temperature of the polymer solution; by adding salt, nonsolvent, incompatible polymer to the polymer solution; or by inducing a polymer-polymer interaction.

Step 2 — Depositing the liquid polymer coating upon the core material. This is accomplished by controlled, physical mixing of the coating material in the manufacturing vehicle. Deposition of the liquid polymer coating around the core material occurs if the polymer is adsorbed at the interface formed between the core material and the liquid vehicle phase, and this adsorption phenomenon is a prerequisite to effective coating. The continued deposition of the coating material is promoted by a reduction

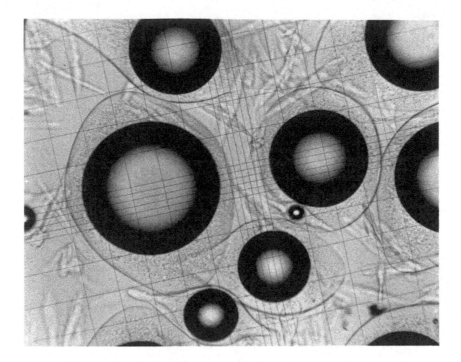

FIGURE 2.   Microencapsulated liquid.

in the total free interfacial energy of the system, brought about by the decreases of the coating material surface area during coalescence of the liquid polymer droplets.

Step 3 — Rigidizing the coating, usually by thermal, crosslinking or desolvation techniques, to form a self-sustaining microcapsule.

III. CORE MATERIAL

A typical microencapsulated liquid and solid are shown in Figures 2 and 3.

Figure 4 shows an encapsulated ferromagnetic material dispersed in an oily liquid. By exposing the capsules to different magnetic fields, the magnetic particles can be aligned in various configurations.

In addition to the single particle structure, aggregate structures can also be produced (Figure 5). The aggregate is composed of a number of core particles in cluster form. The particles in the aggregate need not be the same material and all are individually coated.

In general, water soluble and insoluble solids, water insoluble liquids, solutions and dispersions of solids in liquids can be microencapsulated. A few of the typical materials that have been microencapsulated are shown in Table 1.

IV. PARAMETERS AFFECTING MICROCAPSULE CHARACTERISTICS

The selection of the appropriate coating material dictates, to a major degree, the resultant physical and chemical properties of the microcapsules, and consequently, due consideration must be given this selection. The coating material should be capable of forming a film that is cohesive with the core material; be chemically compatible and nonreactive with the core material; and provide the desired coating properties, such

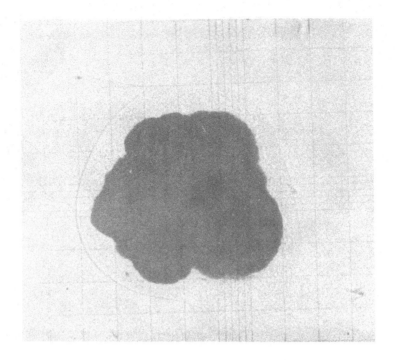

FIGURE 3.   Microencapsulated solid.

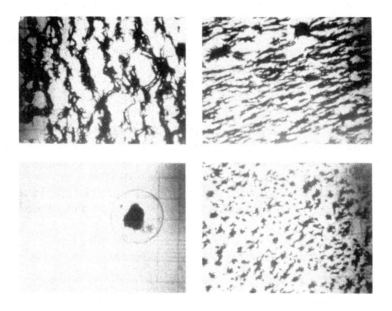

FIGURE 4.   Microencapsulated dispersion.

as strength, flexibility, impermeability, optical properties, and stability. The coating materials used in microencapsulation methods are amenable, to some extent, to *in situ* modification. For example, colorants may be added to achieve product elegance or masking, or coatings may be plasticized or chemically altered through crosslinking,

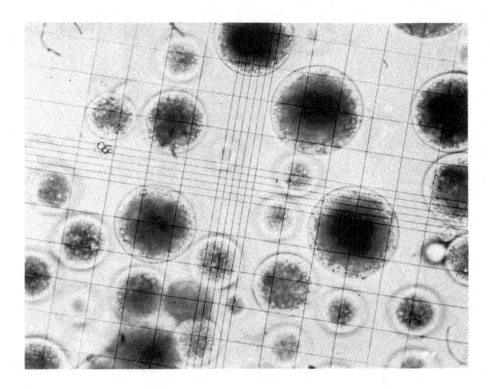

FIGURE 5. Microencapsulated liquid-aggregate.

TABLE 1

Agricultural and Veterinary Core Materials

Analgesics	Fungicides	Nematocides
Anthelmintics	Germicides	Nutrients
Antimicrobials	Growth regulators	Repellents
Bacteria	Herbicides	Pheromones
Disinfectants	Insect diets	Rodenticides
Fertilizers	Insecticides	Virus
Fumigants	Minerals	Vitamins

for instance, to achieve controlled dissolution or permeability. A partial listing of typical coating materials commonly used is suggested in Table 2.

It is not within the scope of this discussion to describe the physical and chemical properties of coatings per se. However, it is pointed out that typical coating properties such as cohesiveness, permeability, moisture sorption, solubility, stability, and clarity must be considered in the selection of the proper microcapsule coating material. The selection of a given coating often can be aided by the review of existing literature and by the study of free or cast films, although practical use of free film information often is impeded for the following reasons:

1. Cast or free films prepared by the usual casting techniques yield films that are considerably thicker than those produced by the microencapsulation of small particles; hence, the results obtained from the cast-films may not be extrapolatable to the thin microcapsule coatings.
2. The particular microencapsulation method employed for the deposition of a

TABLE 2

Typical Coating Materials

Aminoplasts
Carboxymethylcellulose
Cellulose acetate phthalate
Ethyl cellulose
Ethylene vinyl acetate
Gelatin
Gelatin-gum arabic
Gelatin-gum arabic-vinyl
methylether maleic anhydride

Gelatin-gum arabic-ethylene maleic anhydride
Nitrocellulose
Polyvinylalcohol
Propylhydroxycellulose
Shellac
Succinylated gelatin
Saran
Waxes

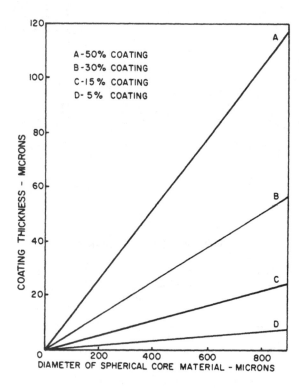

FIGURE 6.   Coating thickness vs. capsule diameter at selected weight percent coatings.

given coating produces inherent properties which are difficult to simulate with
existing filmcasting methods.

3.   The coating substrate or core material may have a decisive effect upon coating
material properties.

Hence, the selection of a particular coating material involves consideration of
both classic free film data and applied results.

The amount of coating can be varied from 1 to 70% by weight. Normal commercial
applications require coatings ranging from 2 to 30%. This corresponds to a dry film
thickness of less than 0.1 to 200 μm, depending on the surface area of the material to
be coated and other physical-chemical parameters of the materials system (Figure 6).

TABLE 3

Approximate Number of Microcapsules per Gram of Material at Various Diameters

Microcapsule diameter (μm)	Approximate number of microcapsules per gram of material (assume density = 1)
5	15,279,000,000
25	122,230,000
50	15,279,000
100	1,909,800
200	238,730
300	70,736
400	29,841
500	15,277
600	8,842
700	5,568
800	3,730
900	2,620
1000	1,910

TABLE 4

Number of Microcapsules per Square Meter at Various Treatment Levels per Hectare and Diameters

Grams of microcapsules/hectares	Approximate number of microcapsules per square meter					
	5 μm	50 μm	100 μm	300 μm	500 μm	800 μm
1	1,527,900	1,530	191	7.1	1.5	0.37
10	15,279,000	15,279	1,910	71	15	3.7
50	76,395,000	76,395	9,550	355	75	19
100	152,790,000	152,790	19,098	707	152	37
200	305,580,000	305,580	38,196	1414	304	74
300	458,370,000	458,370	57,294	2121	456	111
400	611,160,000	611,160	76,392	2828	608	148
500	763,950,000	763,950	95,490	3535	760	185

Microcapsules can be isolated as dry powders or they can be formulated into liquid vehicles capable of being sprayed. If liquid vehicles are desired, a variety of binder-stickers or viscosity builders can be successfully formulated into the final desired product formulation.

Powder microcapsules have been successfully applied with the use of granular applicators, as well as specially designed suction spreaders.

If microcapsules are to be applied in a dispersed liquid state, they can be pumped with lobe, sliding vane, flexible diaphragm or air pressure driven diaphragm pumps. Microcapsules have been successfully sprayed with both single or two fluid nozzles, the diameter of the orifice being several times larger than the largest microcapsule present in the formulation.

Microcapsules range dimensionally in size from a few microns to several thousand microns. Table 3 illustrates the approximate number of microcapsules per gram of material at various diameters.

The number of microcapsules of varying diameters applied per hectare at different application rates is represented in Table 4. Due to the minuteness of the microcapsules, very uniform application rates of the pesticidal materials can be achieved.

Microcapsules can also be compressed or molded into tablets, bricks, sticks, or crayons.

TABLE 5

Stability of Encapsulated Solvents

Encapsulated solvent	Solvent in capsule (%)	Capsule, av. size (μm)	Days on test	Solvent loss at 77°F, 50% RH
Benzene	85.5	500	198	0.5
CarbonTetrachloride	82.8	500	602	0.3
Chloroform	78.9	420	730	0.1
Ethylene Dibromide	66.1	480	730	0.1
Hexane	70.8	35	730	0.1
Toluene	89.4	20	600	0.1
	89.9	90	600	0.1
	90.3	200	400	0.1
	90.6	480	400	0.1
	94.0	720	300	0.1
Trichloroethylene	87.2	500	400	1.0
Perchloroethylene	87.9	500	600	0.1
Xylene	90.2	500	730	0.2

From Herbig, J. A., *Encyclopedia of Chemical Technology*, Vol. 13, 1967, 443.

V. MICROCAPSULE STABILITY

The microencapsulation of materials is of prime importance, but the microcapsules must, in turn, be stable for some period of time; in other words, have an adequate shelf life.

Many volatile liquids can be microencapsulated and subsequently dried to form free-flowing powders. The data presented in Table 5 illustrate the liquid content of microcapsules as a function of time, showing minimal loss for the test conditions.

Microencapsulation cannot yet provide a perfect barrier for materials which degrade in the presence of oxygen, heat, or moisture. The process can, however, often retard the rate of degradation, thereby permitting improved performance. Encapsulation can also provide improved stability to materials affected by a particular formulation component. A typical example of stabilization against environmental influences can be demonstrated by the encapsulation of soybean oil. This liquid can be encapsulated directly and converted to a high active content, powder. The data presented in Figure 7 illustrates the stability enhancement obtained with microencapsulated soybean oil vs. the oil absorbed on a granular carrier.

The stability of controlled release microcapsules can readily be accomplished. Table 6 depicts the stability and release rates of a variety of encapsulated materials under similar laboratory test conditions.

VI. MICROCAPSULE RELEASE MECHANISMS

Release characteristics of microencapsulated materials are another important consideration, and a variety of mechanisms are possible.

The coating can be fractured by external forces, such as pressure, or internal forces as would occur in a microcapsule having a permselective coating. The integrity of the coating can be destroyed by thermal means or by dissolution in an appropriate solvent. In certain instances, release from microcapsules can be accomplished by biodegradative processes.

Release can also be achieved by through-the-wall diffusion. This is especially important for controlled release products. If a water-soluble solid is microencapsulated in a

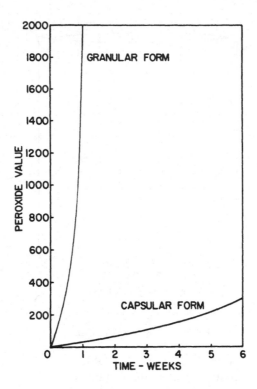

FIGURE 7.   Soil exposure test-encapsulated soybean oil. (From Fogle, M. V., *Croplife*, 1968. With
permission.)

TABLE 6

**Stability of Controlled Release Patterns of Solids Having Varying
Degrees of Water Permeability**

| | | % Release | | |
Material	Time (hr)	Initial	Final	Time (months)
Acetoaminophen	1	18	16	21
	2	35	32	—
	3	49	46	—
	5	69	67	—
Ammonium dichromate	1	76	72	10
Acetylsalicylic acid	1	16	16	23
	2	30	30	—
	3	44	45	—
Sodium bicarbonate	1	9	8	30
	2	14	14	—
	4	30	30	—
Potassium chloride	1	74	76	25
	2	96	96	—

water-insoluble film, such as ethylcellulose, the content of the microcapsule can be
extracted with water. A simplified release mechanism is depicted in Figure 8. In the
initial stage of the process, water permeates the coating (R1). Next, an aqueous solution of the water soluble solid is formed within the structure (R2) and this solution in

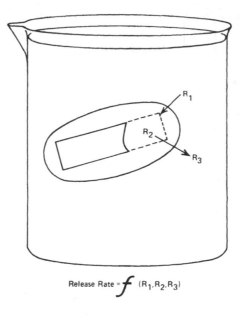

Release Rate = f (R_1, R_2, R_3)

FIGURE 8.   Release by diffusion.

turn permeates into the continuous water phase (R3). Hence, the release rate is a function of the film's permeability to water, the solubility of the microencapsulated solid, and the permeability of the film to the saturated solution. This release mechanism is independent of acidic or basic conditions, provided the film structure and the solubility of the microencapsulated material are insensitive to the varied pH conditions. The resultant release rate can usually be described as a psuedo-first order rate process described by the following equation -dc/dt = kc where c is the concentration of material remaining in the capsule, t is time, and k is the first order velocity constant (Figure 9). It has been found that for a given microcapsular system, the velocity constant k can be very reproducible.

Effective control of diffusion release properties can be achieved with materials having diverse water solubilities. Judicious selection of an appropriate coating material and variation of the coating thickness allow a spectrum of patterns to be achieved.

VII. PRODUCT CONSIDERATIONS AND RESULTS

Microencapsulation processes can be effectively used to convert liquids to solids, separate reactive materials, reduce material toxicity, provide environmental protection to compounds or formulations, alter surface properties, control the release of materials, reduce the volatility or flammability of liquids, and for taste-masking bitter compounds. These are attributes more commonly derived from the microencapsulation of materials. However, they should not be considered the only contributions microcapsules can make to agricultural products. The following summarily elaborates on some of these contributions.

A. Converting Liquids to Solids

Microencapsulation can be used to effectively convert liquids to solids. As an illustrative example, Mirex, a fireant (*Solenopis saevissima ichteri*) toxicant has been encapsulated as a soybean oil solution. The dry microcapsules were designed to allow

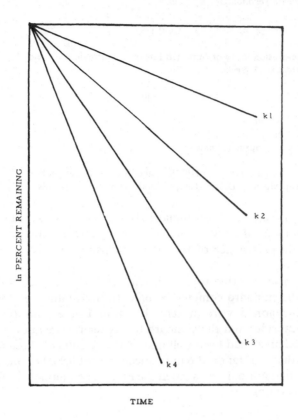

FIGURE 9. Typical microcapsule release rates from ethylcellulose capsules.

TABLE 7

Effectiveness of Fireant Bait

November 1967	Percent mound mortality 22 weeks
Standard bait	96
Microencapsulated bait	97
May 1968	
Standard bait	92
Microencapsulated bait	94
August 1968	
Standard bait	100
Microencapsulated bait	100
February 1969	
Standard bait	61
Microencapsulated bait	100

Note: Plot size: 1 acre in triplicate.

From Markin et al.[11,12,13]

ants to carry the microcapsules back to their mound and feed on the contents, thus infesting the colony.

Table 7 shows equivalent mortality of microencapsulated bait compared to a standard commercial form marketed at that time. The microcapsules were disseminated at

TABLE 8

Reinvasion Rates of Imported Fireant into Previously Cleared
Areas at 15 Weeks

	Mounds/acre (av 3 plots each)
Check plots	94.0
Standard bait	9.0
Microencapsulated bait	0.67

From Markin, G. and Hill, S., paper presented Entomol. Soc.
Am. Meeting, Dallas, Tex., December 1968. With permission.

100 g/acre vs. 565 g/acre for the standard commercial product. Both formulations applied the same amount of active toxicant per acre. The investigators noted a significant increase in field persistence of the microencapsulated bait over the commercial standard.

Markin and Hill also reported that in the fall of 1967, several thousand acres had been treated with the standard commercial bait. In the summer of 1968, it was noted that no old fireant colonies were present. Random 1-acre plots (in triplicate) were treated with microcapsules and the standard bait after reinvasion of the queens from surrounding infested areas had been well underway. Results of these tests are in Table 8. They concluded that the standard bait eradicated established mounds. However, the microencapsulated bait not only eradicated the established mounds, but also prevented new mounds from forming.

B. Increased Flowability

Because a microcapsule coating can change particle geometry, alter moisture sorption, add lubricity, and change particle surface properties, microencapsulated materials can be made to flow more readily than unencapsulated materials.

Hygroscopic materials characteristically tend to absorb water vapor prematurely during storage and handling. Excessive clumping and caking results, thereby destroying the original particle size spectrum and the handling properties. Microencapsulation significantly reduces the clumping and caking that occurs with bulk hygroscopic powders.

Figure 10 shows bulk powders of unencapsulated and microencapsulated urea exposed to a humid atmosphere for 2 hr, then passed through a 100-mesh screen; 90% of the unencapsulated urea was retained on the screen, whereas, only 9% of the microencapsulated urea was retained.

This property could be effectively used with agricultural chemicals and has been used with cloud seeding materials.

C. Reduced Toxicity

Microencapsulation can also be used to control the bioavailability of materials. In selected instances, this control can cause a reduction in toxicity of the materials. For example, in a test conducted to control the level of bioavailability of acetylsalicylic acid, the LD_{50} of unencapsulated acetylsalicylic acid in rats was found to be 1750 mg/kg, whereas, the LD_{50} of a microencapsulated form of acetylsalicylic acid was found to be 2823 mg/kg.

This contribution of microcapsules should be very helpful for formulating and disseminating agricultural materials.

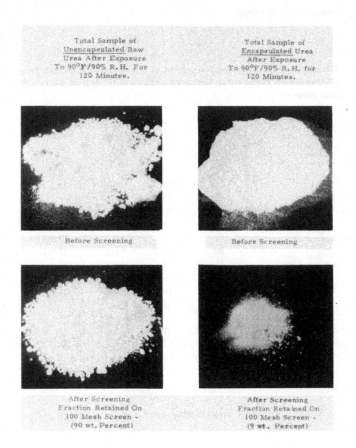

FIGURE 10. Clumping tendency on unencapsulated and microencapsulated urea.

D. Controlled Release

Controlled release can, of course, be viewed from many aspects. Researchers, like Miller and Cordon,[14,15] have used microencapsulation for controlling the release, or possibly better stated, inhibiting the release of pesticides so that they may pass through cows and yet control flies in fecal matter.

They reported work with NCR microencapsulated *stirofos* for control of the housefly, (*Musca domestica L.*) when fed to lactating dairy cows. They found that in an initial insecticidal balance experiment that 15% of the microencapsulated *stirofos* reached the feces as compared to 0.3% found with a 75% wettable powder formulation. In essence, no *stirofos* appeared in the milk.

Confirmatory trials indicated that microencapsulation did allow more *stirofos* to pass through the digestive tract of the bovine, thus increasing the fecal pesticide concentration evidenced by Table 9.

Larval mortality was significantly better with the microencapsulated formulations over the control, whereas, differences in pupal mortality were insignificant.

Yet another application of encapsulation of controlled release involves pheromones for pest population management. Beroza et al.[16-19] have demonstrated the real possibilities of microencapsulated disparlure for control of the gypsy moth (*Portheteria Dispar L.*).

Table 10 was abstracted from one of Beroza's papers indicating, in part, the type of control (measured by moth capture techniques) attainable via microencapsulation.

In later experiments it was shown by egg-mass counts that substantial suppression

TABLE 9

Rabon Fecal Residues from Cows Fed Two Levels of 5 Formulations of Rabon and Larval
Mortalities, Pupal Weights, and Pupal Mortalities of Beltsville Resistant Strain Fly Larvae
Seeded into the Feces

Treatment	Level	Fecal reside (ppm)	Larval mortality (%)	Pupal weight (mg)	Pupal mortality (%)
Control	0	0.09	9.7	26.6	7.5
75% WP	15	0.16	27.5	25.6	7.3
	30	0.22	64.2	24.1	12.0
AC-357 microcapsules	15	2.94	48.5	24.2	7.9
	30	6.17	80.5	22.7	14.0
AC-358 microcapsules	15	4.17	77.6	23.0	8.2
	30	7.38	95.3	22.3	12.0
AC-359 microcapsules	15	4.21	58.8	23.5	10.2
	30	6.38	95.8	20.9	15.6
AC-360 microcapsules	15	2.36	91.7	21.1	3.3
	30	6.03	100.0	—	—

(From Miller, R. W., Drazek, P. A., Martin, M. S., and Gordon, C. H., Feeding of
Microencapsulated Gardona for the Control of Fly Larvae in Cow Manure, paper presented Am. Dairy Sci. Assoc. Meeting, Gainesville, Fla., June 1970. With permission.)

TABLE 10

Effectiveness of Microencapsulated Disparlure of Captures of Released Gypsy Moth
Males in Monitor Traps

Treatment (lure/ha)	Release number	Days post-treatment	Number males released	Percent suppression	Total males captured
Untreated	1	—	240	—	1
	2	—	675	—	89
	3	—	784	—	304
	4	—	390	—	144
	5	—	387	—	138
Capsules (1.8 g/ha)	1	9	245	—	0
	2	23	618	93	6
	3	30	772	95	14
	4	37	378	99	1
	5	44	385	93	10
20/40 cork (11.1 g/ha)	1	9	190	—	0
	2	23	661	94	5
	3	30	736	100	1
	4	37	375	99	1
	5	44	385	91	12

From Beroza, M., Steven, L., Bierl, B., Philips, F., and Tardif, J. G. R., *Environ.
Entomol.*, 2(6), 1051, 1973. With permission.

of mating can result from treatment with microencapsulated disparlure as indicated in
Table 10.

To summarize, the field life of disparlure can be extended to approximately 6 to 8
weeks. This type microcapsule principle is presently being expanded to numerous other
Lepidoptera insect pheromones and other successes are being reported in the literature.

TABLE 11

Mean Egg-Mass Counts Per Plot ± SE in 100 0.1-ha (0.25-acre) Plots in Untreated Area and in Area Treated with Disparlure Microcapsules (5 g lure/ha)

Area	Mean prespray counts	Mean postspray counts	
		Fertile	Infertile
		Original plots	
Microcapsules	0.92 ± 0.09	1.47 ± 0.23 (160%)[a]	0.25 (15%)[b]
Untreated control	1.49 ± 0.15	5.27 ± 0.72 (354%)[a]	0.10 (1.9%)[b]
		Subplots	
Microcapsules		1.25 ± 0.20 (136%)[a]	0.25 (17%)[b]
Untreated control		4.37 ± (293%)[a]	0.09 (2.0%)[b]

[a] Percent of prespray count = prespray count/postspray count × 100

[b] Percent of total no. egg masses that were infertile = infertile egg masses/fertile + infertile egg masses .

From Beroza, M., Hood, C. S., Trefrey, D., Leonard, D. E., Knipling, E. F., Klassen, W., and Stevens, L. J., *J. Econ. Entomol.*, 67(5), 659, 1974. With permission.

Turning our attention to encapsulated organophosphate insecticides, we find that controlled release or increased field persistence can also be obtained with this type chemical moiety. Figure 11 represents the variation in release rates of various malathion microcapsules in aqueous buffer solution using a rotating bottle apparatus.

Figure 12 depicts the corresponding in-soil persistence of malathion microcapsules as measured in laboratory soil tests. This type of developmental product should merit further field test considerations.

E. Taste Masking

Controlled release can be viewed as operating over a period of minutes as opposed to hours, days, or months, if the release of materials can be achieved without appreciably altering the bioavailability.

Figure 13 displays a taste-masked presentation of a microencapsulated antiinflammatory agent phenylbutazone marketed for horses in various parts of the world.

In addition to the veterinary medicinal field, microencapsulation should be considered as a potentially valuable tool for taste-masking rodenticides.

F. Microbial Materials

Microbial materials can also be microencapsulated. A protective coating can potentially extend their activity, stabilize the material, and protect the user from the toxicant.

Raun and Jackson[20] have reported that *Bacillus thuringiensis* has been successfully microencapsulated, as evidenced by Table 12. The microcapsular material retained its viability and pathogenicity for the European corn borer, (*Ostrinia nubilalis* [Hubner]), as evidenced by Table 13 and 14.

In field tests the microcapsule formulation gave good control as can be seen in Tables 15 and 16.

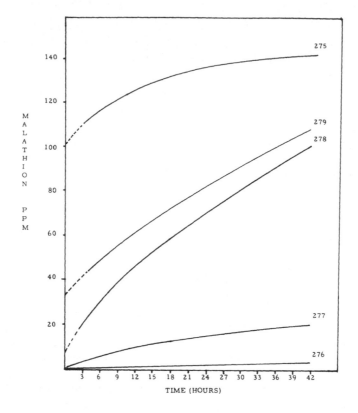

FIGURE 11.   Release of malathion in aqueous buffer solution.

VIII. ADVANTAGES AND LIMITATIONS OF SYSTEMS

This technology is not exclusive of problem areas. For instance, no single process is adaptable to all core material candidates or product applications. Difficulties, such as incomplete or discontinuous coating, inadequate stability or shelflife of sensitive, non-reproducible and unstable release characteristics of coated products, and economic limitations, often are encountered in the attempt to apply a particular microencapsulation method to a specific task. Many times, successful adaptation is, in part, a result of the technical ingenuity of the investigators.

IX. LABORATORY AND MANUFACTURING EQUIPMENT

The coacervation/phase separation methods are basically batch type processes. In the laboratory, beakers equipped with variable speed agitation are used to prepare microcapsules (Figure 14). In pilot plant and production operations, the microcapsules are produced in water-jacketed stainless steel tanks equipped with variable speed agitators (Figures 15 and 16). Solution preparation and core material holding tanks are normally used for preprocessing of materials. If the microcapsules are to be concentrated or dried, then this type of equipment is also required for subsequent processing.

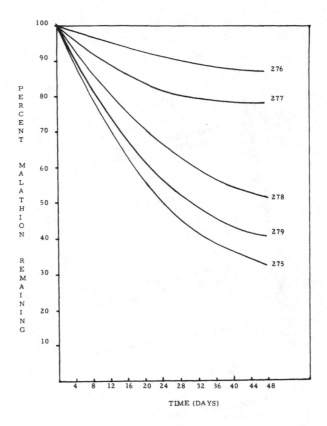

FIGURE 12.   In-soil (laboratory) persistence.

TABLE 12

Encapsulated *Bacillus Thuringiensis* Formulations Received at the European
Corn Borer Research Laboratory in August 1963

Sample designation	Percent *B. Thuringiensis*	*B. Thuringiensis* Spore/g of formulation[a] ($\times 10^8$)
Microcapsules LDM 3	97.5	1560
Microcapsules LDM 4	26.0	416
Attapulgus clay granule	—	200
Spore powder (control)	—	1600

[a] Estimates based on actual counts of preformulated powder and formulator's
statement of contents.

From Raun, E. and Jackson, R., *J. Econ. Entomol.*, 59(3), 620, 1966. With
permission.

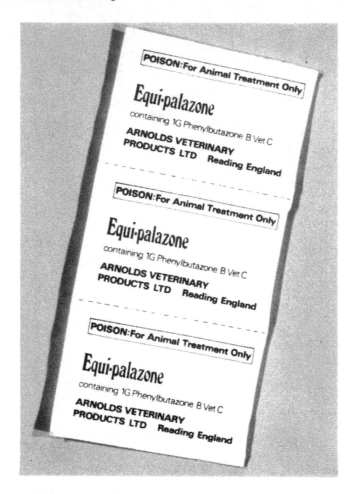

FIGURE 13. Taste-masked phenylbutazone. (Courtesy Arnolds
Veterinary Products, Ltd., Reading, England)

TABLE 13

Mortality of European Corn Larvae Fed Green Beans Soaked for 19 Min in a Suspension of *Bacillus Thuringiensis*, Ankeny, Iowa, 1964

Formulation	Dosage in spores/0.1 cc ($\times 10^6$)	Mortality			
		48 hr		72 hr	
		%	95% C.l.[a]	%	95% C.l.[a]
Microcapsules LDM3	15.6	75	59-87	85	70-94
	3.9	30	17-47	40	25-57
	0.95	30	17-47	30	17-47
Microcapsules LDM 4	15.4	60	43-75	85	70-94
	3.0	15	6-30	45	29-62
	0.95	5	1-17	40	25-57
Spore powder (control)	15.2	60	43-75	90	76-97
	3.8	30	17-47	40	25-57
	0.95	20	9-36	20	9-36
Attapulgus clay granule	15.2	60	43-75	90	76-97
	3.8	40	25-57	40	25-57
	0.95	10	3-24	20	9-36
Check (water only)	0.0	0	0-9	0	0-9

[a] Confidence interval.

From Raun, E. and Jackson, R., *J. Econ. Entomol.*, 59(3), 620, 1966. With permission.

TABLE 14

Mortality of European Corn Borer Larvae Fed Green Beans Soaked for 19 Min in a Suspension of *Bacillus Thuringiensis*, Ankeny, Iowa, 1965

Formulation	Viable spores/g ($\times 10^9$)	Dosage in spores/0.1 cc ($\times 10^6$)	%	95% C.l.b	%	95% C.l.a
Microcapsules LDM-3	156	15.6	52	36-70	52	—
	—	3.9	32	19-50	32	—
	—	0.98	8	2-21	8	—
Microcapsules 1965 LDM-4	110[a]	15.6	76	61-80	84	69-93
	—	3.0	80	64-91	84	69-93
	—	0.08	44	20-62	56	40-73
Spore powder (control)	100	15.6	64	48-79	80	64-91
	—	3.9	56	40-73	64	47-79
	—	0.98	32	19-50	48	33-66
Check	—	0.0	8	2-21	8	—

[a] Confidence intervals.

From Raun, E. and Jackson, R., *J. Econ. Entomol.*, 59(3), 620, 1966. With permission.

TABLE 15

Effectiveness of Granular and Capsule Formulations of *Bacillus Thuringiensis* in the Control of First Generation European Corn Borer Larvae, Ankeny, Iowa, 1964

Formulation	Dosage/ acre (lb)	Spore equiva- lent/acre[a]	% Control over check[b]
Microcapsules LDM-3	4	31.2	82.1
	3	23.4	62.1
Microcapsules LDM-4	4	11	44.7
Attapulgus clay granule	20	20	83.5
	15	15	70.5
5% DDT (granular)	20		79.0

[a] 1 spore equivalent = 1 lb (454 g) material containing 20×10^9 spores.

[b] Differences between capsular and granular formulations were not significant. Differences within capsular formulations were not significant except that the 4 lb dosage of LDM-4 differed at the 5% level from the 4 lb dosage of LDM-3.

From Raun, E. and Jackson, R., *J. Econ. Entomol.*, 59(3), 620, 1966. With permission.

TABLE 16

Effectivemess of Granule, Capsule, and Spray Formulations of *Bacillus Thuringiensis* in the Control of First Generation European Corn Borer Larvae, Ankeny, Iowa, 1965

Application method	Formulation des- ignation	Field dos- age/acre	Spore equiva- lents/acre[a]	% Control[b]
Granular applicator	Microcapsules			
	LDM-3	4 lb	31	85.1 c
	1965 LDM-4	2 lb	11	63.8 d
		3 lb	16.5	40.4 d
		4 lb	22	70.2 c
	Attapulgus clay granules	20 lbs	20	89.4 c
	5% DDT	20 lbs	—	76.6 c
Sprayer	DDT	1.5 lb		90.1 c
	Microcapsules			
	1965 LDM-4	1.5 lb	8.3	90.1 c
		4.0 lb	22	74.0 c

[a] 1 spore equivalent = 1 lb (454 g) material containing 20×10^9 spores.

[b] Results followed by the same letter designation do not differ significantly from one another at the 0.05 level, Duncan's multiple range test.

From Raun, E. and Jackson, R., *J. Econ. Entomol.*, 59(3), 620, 1966. With permission.

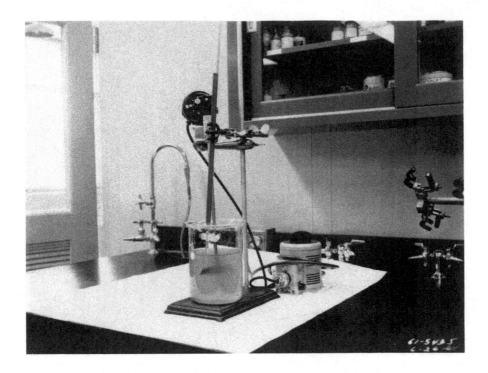

FIGURE 14.   Laboratory microencapsulation equipment.

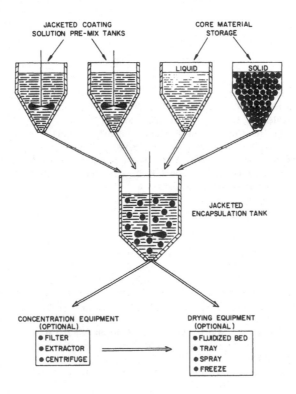

FIGURE 15.   Microcapsule flow diagram.

FIGURE 16.   Microcapsule pilot plant.

REFERENCES

1. **Green, B. K.,** U.S. Patent 2,712,507, 1955.
2. **Green, B. K. and Schleicher, L.,** U.S. Patent 2,730,456, 1956.
3. **Green, B. K. and Schleicher, L.,** U.S. Patent 2,800,457, 1957.
4. The National Cash Register Corporation, British Patent 907,284, 1963.
5. **Miller, R. E. and Anderson, J. L.,** U.S. Patent 3,155,590, 1964.
6. **Miller, R. E., Fanger, G. O., and McNiff, R. G.,** South African Patent 4211-66, 1967.
7. **Heistand, E. N., Wagner, J. G., and Knoechel, E. L.,** U.S. Patent 3,242,051, 1966.
8. **Green, B. K.,** U.S. Patent Reissue 24,899, 1960.
9. **Brynko, C., Bakan, J. A., Miller, R. E., and Scarpelli, J. A.,** U.S. Patent 3,341,466, 1967.
10. **Fogle, M. V.,** Microencapsulation promises big potential in agricultural field, *Croplife,* February 1968.
11. **Markin, G. and Hill, S.,** Microencapsulated Oil Bait for the Control of the Imported Fire Ant, paper presented Entomol. Soc. Am. Meeting, Dallas, Tex., December 1968.
12. **Markin, G., Mauffray, C., and Adams, D.,** A granular applicator for very low volumes of microencapsulated insect bait or other materials, Agricultural Research Service, U.S. Department of Agriculture, Beltsville, Md., October 1969.
13. **Markin, G. and Hill, S.,** Microencapsulated oil bait for control of the imported fire ant, *J. Econ. Entomol.,* 61, 193, 1971.
14. **Miller, R. W., Drazek, P. A., Martin, M. S., and Gordon, C. H.,** Feeding of Microencapsulated Gardona for the Control of Fly Larvae in Cow Manure, paper presented Dairy Sci. Assoc. Meeting, Gainesville, Fla., June 1970.
15. **Miller, R. W. and Gordon, C. H.,** Encapsulated rabon for larval house fly control in manure, *J. Econ. Entomol.,* 65(2), 455, 1972.
16. **Beroza, M., Steven, L., Bierl, B., Philips, F., and Tardif, J. G. R.,** Pre- and post-season field tests with disparlure, the sex pheremone of the gypsy moth, to prevent mating, *Environ. Entomol.,* 2(6), 1051, 1973.

17. Beroza, M., Hood, C. S., Trefrey, D., Leonard, D. E., Knipling, E. F., Klassen, W., and Stevens, L. J., Large field trial with microencapsulated sex pheromone to prevent mating of the gypsy moth, *J. Econ. Entomol.*, 67(5), 659, 1974.

18. Gentry, C. R., Beroza, M., Blythe, J. L., and Bierl, B. A., Efficacy trials with the pheromone of the oriental fruit moth and data on the lesser appleworm, *J. Econ. Entomol.*, 67(5), 607, 1974.

19. Cameron, E. A., Schwalbe, C. P., Stevens, L. J., and Beroza, M., Field tests of the olefin precursor of disparlure for suppression of mating in the gypsy moth, *J. Econ. Entomol.*, 68(2), 158, 1975.

20. Raun, E. and Jackson, R., Encapsulation as a technique for formulating microbial and chemical insecticides, *J. Econ. Entomol.*, 59(3), 620, 1966.

Chapter 5

PHARMACEUTICAL APPLICATIONS OF MICROENCAPSULATION USING COACERVATION/PHASE SEPARATION TECHNIQUES

M. Calanchi and M. Maccari

TABLE OF CONTENTS

I. INTRODUCTION

Both in human and in veterinary medicine, the techniques of microencapsulation described in Chapter 4 find diversified applications, several of which have already been tested clinically and approved by health authorities in many countries where microencapsulated drug products have gained excellent professional and public acceptance. In essence, microencapsulation of drugs intended for ingestion affords a number of distinct (if sometimes overlapping) effects; the most important are the conversion of liquids to pseudo-solids, sustained release, better gastrointestinal tolerability, taste abatement, and separation of mutually reactive components. These are discussed in this chapter.

II. CONVERSION OF LIQUIDS TO PSEUDO-SOLIDS

Drugs that occur naturally in a liquid form can be converted by microencapsulation to a pseudo-solid, free-flowing powder which is suitable for compounding with other active substances, tableting, filling into standard hard gelatin capsules, and other dosage forms that cannot be made with liquids. Examples of such drugs are cod liver oil, castor oil, clofibrate, medium-chain triglycerides, vitamin E, and many others. We'll take clofibrate for illustration. Clofibrate is a drug widely used for the treatment of patients with hypercholesterolemia (too much cholesterol in the blood) and hyperlipemia (too much fat in the blood). It occurs as an oily liquid with a most unpleasant taste and odor, so that its use in the natural form is practically impossible—more so since the dosages required are rather large. Microencapsulation, however, converts clofibrate to a dry, free-flowing powder, consisting of microscopic droplets of uniform size, individually coated with an ultrathin membrane made of a pharmacologically inert and completely harmless material. This pseudo-powder behaves exactly like any other powder and can be filled into hard gelatin capsules with standard capsule-filling machinery (Figure 1).

In some detail: the clofibrate pseudo-powder contains between 75 and 90% active core material and between 10 and 25% membrane material (gelatin); at least 95% of the microencapsulated droplets are less than 500 μm in diameter. The active material is liberated from the ingested capsule by enzymatic digestion of the coating material; a simulated gastric and intestinal juice is used for testing of the microencapsulated product in the quality control laboratory.

In addition to converting the clofibrate oil to a free-flowing powder, the continuous membrane coating applied on each tiny droplet of the substance also abates its taste and odor— a bonus of no mean order, and it protects the active substance from external injury (increase stability). These aspects are discussed in greater detail later.

III. SUSTAINED RELEASE

With drugs that are sufficiently soluble at physiological pH values, the coating membrane of microcapsules can be engineered so that it will provide sustained release of the active content by continuous dissolution (see also Chapter 4). This, in turn, affords several advantages, namely:

1.   Reproducible control of release rate and in some cases, control of the site of release
2.   Materially better drug utilization
3.   Markedly better gastrointestinal tolerability

FIGURE 1. Microencapsulated clofibrate (oil) demonstrating flowability. (Photograph courtesy of Medicamehta Company, Lisbon.)

4. More uniform drug concentrations in the blood without unduly high "peaks" such as are often responsible for adverse effects
5. Greater patient convenience (and compliance) with more comfortably spaced doses

Many drugs have been successfully microencapsulated to obtain sustained release with all its inherent advantages; the following example is representative enough to account for the whole category.

Acetylsalicylic acid (aspirin) is probably the best known drug ever to appear in microencapsulated dosage forms.[1-4] Among several published studies on microencapsulated aspirin, the one by Saggers et al.[1] was designed with beautiful simplicity, yielded simple and straightforward answers, and so appears particularly suitable for a brief description. The experiment involved six subjects and was conducted in the classical cross-over fashion with a 1-week washout period between doses. Each subject took 1000 mg of aspirin, once as raw acetylsalicylic acid and once as tablets made from microencapsulated acetylsalicylic acid; blood samples were obtained from each partic-

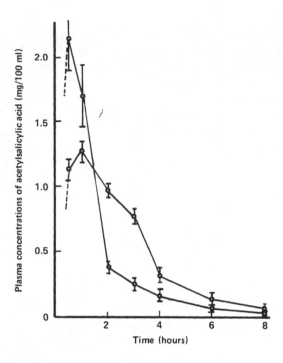

FIGURE 2.   Plasma concentration-time relationship
of acetylsalicylate after administration of 1 g aspirin in
a sustained-release preparation (•—•), and 1 g aspirin
in a standard tablet (o—o). Each point is the mean, ±
SEM, of observation on six subjects. (Reprinted from
Saggers, V. H., Chasseaud, L. F., and Cooper, A. J.,
Clin. Trials J., 9(3), 37, 1972. With permission.)

ipant at suitable intervals during 8 hr after dosing, and all samples were assayed for
unchanged aspirin and its principal metabolite, salicylic acid. Figure 2 clearly indicates
that the conventional tablet form of the drug gave materially higher plasma concentrations of aspirin when first assayed — in fact, the peak level in plasma was very nearly
double that obtained with the sustained-release preparation, and the peak itself occurred somewhat earlier; but after completion of the initial absorptive phase, aspirin
was maintained at higher levels after administration of the microencapsulated formulation. This pattern of aspirin blood levels suggests that analgesia from the microencapsulated dosage form would be more prolonged than with the same amount of aspirin taken as an ordinary tablet. Another interesting or indeed fundamental
observation emerging from this study is that in terms of bioavailability, calculated
from the areas under the concentration-time relationship curves, aspirin was equally
available from both formulations. In other words, while the microencapsulated drug
exited from the microcapsules at a deliberately slow rate, all of the active content was
released within 8 hr of ingestion.

Other clinical investigators paid special attention to the improved gastrointestinal
tolerability of the microencapsulated aspirin compared to the raw drug. In two studies
dealing particularly with this aspect,[3,4] the sustained-release product proved materially
superior.

A typical batch of microencapsulated aspirin contains between 92 and 98% active
core material and between 1.3 and 7.7% capsule wall material (in this instance, ethylcellulose); at least 95% of the microcapsules are less than 870 μm in diameter. The

Table 1

Daily Fecal Blood Loss (ml) After Oral Administration of Sustained Release Microencapsulated KCl (3g/day) in 20 Volunteers.

Ml of blood/24 hr.	Days of treatment				
	0	10	30	60	90
Mean	0.576	0.516	0.432	0.456	0.437
S.D.	±0.046	±0.047	±0.048	±0.048	±0.052

Modified from Maggi, G. C. and Coppi, G., *Curr. Ther. Res. Clin. Exp.*, 21 (5), 676, 1977. With permission.

encapsulated material will release between 40 and 60% of its active content when agitated for 1 hr in a simulated gastric juice at pH 1.5 and the rest gradually over the next 7 hr in simulated intestinal juices. The tolerated quota of free (split) salicylic acid is less than 0.3%.

IV. AVOID GASTROINTESTINAL IRRITATION

Many drugs that are taken orally irritate the gastrointestinal mucosa; some will literally make holes in the stomach, i.e., produce ulcerations. Even without going to such extremes, gastrointestinal irritation is distressing to the patient and interferes seriously with proper absorption of the administered drug. Obviously, this irritating action is greater the higher the local concentration of the irritating chemical (drug) on or near the mucosa. Now, since a drug microencapsulated for sustained release cannot by definition come out all at once from the dialyzing microcapsules, any drug so processed will invariably be tolerated materially better than the same amount of the same drug in the unprocessed form. This has proved so true in general clinical use that a special microencapsulation technology was evolved to secure the same kind of protection from gastric irritation with drugs that do not, strictly speaking, need to be processed for sustained release. In such a case, indeed, the preoccupation is in a way the opposite — namely to make sure that the microencapsulated drug retains its full measure of bioavailability: this is achieved by careful adjustment of the amount of coating material used.

A very good example is the microencapsulation of potassium chloride — a simple, but quite irritating, chemical. Potassium salt is used therapeutically to prevent, minimize, or correct hypokalemia (potassium depletion in the blood). The finished product contains approximately 85% active core material (potassium chloride crystals) and 15% membrane wall material (cellulose polymer), with 95% of the capsules less than 870 μm in size.

In one human clinical study,[5] the gastrointestinal tolerability of microencapsulated potassium chloride was assessed in terms of daily blood losses into the feces in 20 patients treated with the microencapsulated product in the amount of 3 g daily for 3 consecutive months (fecal blood losses were measured with a high-precision method involving the labeling of the patient's own blood cells with ^{51}Cr). As shown in Table 1, the mean daily blood losses never exceeded pretreatment control values, ranging as they did between 0.432 and 0.516 ml; in contrast, patients receiving raw potassium chloride were reported to lose up to 2.2 ml daily.

Another clinical trial[6] was designed to demonstrate the bioavailability of potassium chloride from the microencapsulated dosage form when the product was administered to patients with potassium depletion induced by the use of diuretics. Because metabolic

TABLE 2

Effectiveness of Sustained Release Microencapsulated KCl in Six Patients with Hypokalemia in Terms of Kühns' KTD. (see text)

	Upon discontinuation of diuretic therapy	On third day off diuretics	(1st KTD test) immediately after first potassium load	(2nd KTD test) after 7 consecutive days on sustained release microencapsulated KCl
K + Serum/mEq/l				
Mean	3.20	3.25	3.48	4.00
S.D.	±0.24	±0.22	±0.20	±0.23*
K + Excretion mEq/day				
Mean	41.7	37.2	42.3	61.3
S.D.	±8.0	±7.7	±10.2	±17.2*

* Significant with Student "t" test for paired data ($P \leqslant 0.01$) first KTD.

Modified from Maggi, G. C. and Coppi, G., *Curr. Ther. Res. Clin. Exp.*, 21(5), 676, 1977. With permission.

balance studies with calculation of total potassium intake from food and fluid consumed and of potassium excretion into the urine and feces are inordinately laborious, the authors adopted a simpler method based on Kühns' "oral test for potassium deficiency", abbreviated to "KDT" from the original German initials. This involves administration of an oral potassium load and a follow-up study of its urinary excretion — whose increase obviously reflects the fact that the potassium salt taken orally was to that extent absorbed. The trial was conducted as follows. Six patients with demonstrated hypokalemia secondary to diuretic medication (dichlorophenamide plus hydrochlorothiazide) were taken off said medication and given an oral load of potassium chloride in the microencapsulated dosage form in the amount of 4 capsules daily (representing 32 mEq of K⁺), which was then repeated for 7 consecutive days. Based on the results tabulated in Table 2, the following remarks can be made:

1.   All six patients had definite hypokalemia from chronic diuretic medication; basal blood potassium assays on the first and third day off diuretics being, respectively, 3.2±0.24 mEq/l. Urinary potassium was rather on the low side, in keeping with a situation of hypokalemia.

2.   The first Kühns test, administered immediately after the initial potassium load, revealed no significant modification of blood potassium assays, the mean being 3.48±2.0 mEq/l; urinary excretion remained essentially unchanged.

3.   At the end of the 7-day treatment, however, there was a highly significant increase of mean blood potassium assay (to 4.0±0.23 mEq/l and a likewise highly significant increase of mean urinary potassium content (to 61.3±17.2 mEq in 24 hr as opposed to only 42.3±10.2 mEq at the start of treatment).

These results clearly indicate that the potassium chloride contained in the microencapsulated dosage form was readily available from the medication and in fact thoroughly absorbed.

V. TASTE ABATEMENT

One step short of the application just discussed is the microencapsulation of drugs for the purpose of abating objectionable taste and odor (commonly, if properly, called "taste masking") in drugs for oral administration, particularly in pediatric dosage forms. Here the idea is to prevent contact between an unpalatable (mostly bitter) drug and the taste buds while the dose traverses the mouth and pharynx. Conversely, once in the stomach, the drug is supposed to be immediately (or nearly so) available for absorption; in other words, taste abatement does not imply sustained release, even though sustained release does imply at least a measure of taste abatement. In terms of quality control, taste abatement is judged only empirically, whereas bioavailability must be demonstrated by rigorous scientific methods.

The drugs more commonly microencapsulated for taste abatement are vitamins and antibiotics; among the latter, an excellent example is the semisynthetic penicillin congener, flucloxacillin. One of us[7] conducted a special study to assess the bioavailability of a flucloxacillin preparation microencapsulated for taste abatement with 17% ethylcellulose, with an in vitro release of 80% in half an hour, made up as a granular product for extemporaneous resuspension. The reference product was a commercially available flucloxacillin preparation which was repacked so as to be indistinguishable from the test product and so afford "double blind" experimental conditions. The test subjects were six healthy adult males maintained in uniform housing and feeding conditions throughout the test period. The experimental design was the classical cross-over pattern with two-by-two Latin square distribution, whereby each subject received both products on two separate occasions with a suitable washout period between dosings (in other words, each subject served as his own control). On each test day, blood samples were obtained from each participant at 30 min and at 1, 2, 4, and 6 hr after dosing to be assayed for flucloxacillin content. Also, urinary excretion of the antibiotic was measured at 6 and 24 hr after dosing. Figure 3 shows the plasma levels of flucloxacillin elicited at the various sampling times with the test product and with the raw reference material. Clearly, both dosage forms were readily available to absorption and produced much the same peaks and general curve pattern; indeed, if you measure the areas under the time/concentration curves, you find that the microencapsulated product was slightly more "available" than the reference product, though the difference was not statistically significant. These results are confirmed by the data of urinary excretion shown in Figure 4, again indicating a small (not significant) difference in favor of the microencapsulated product. This trial proves that flucloxacillin microencapsulated for taste abatement is fully as available from the dosage form as the raw (unprocessed) antibiotic.

VI. SEPARATION OF INCOMPATIBLES

Mutually reactive materials going into a compound formulation can be effectively prevented from interacting by microencapsulating one or more of them as the case may require. A typical example is aminopropylon which is microencapsulated with a cellulose polymer for compounding with thiamine, pyridoxine, and the cobalamins.[8] Of course, there are other ways to separate incompatibles in a compound formulation — one being the classical "multilayer tablet", but this is far more laborious to make and performs no better than microencapsulation in terms of stability of the formulation. Figure 5 shows the results of an accelerated stability test involving three dosage forms of the same compound formulation, namely an ordinary tablet, a three-layer tablet, and a tablet made with microencapsulated aminopropylon; stability was meas-

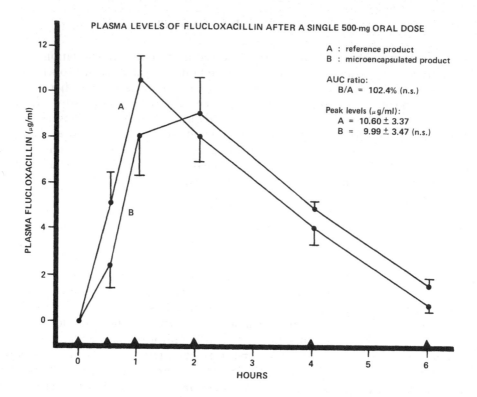

FIGURE 3.   Plasma levels of flucloxacillin after a single 500-mg oral dose.

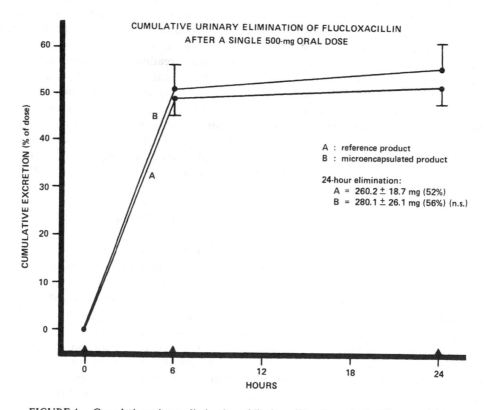

FIGURE 4.   Cumulative urinary elimination of flucloxacillin after a single 500-mg oral dose.

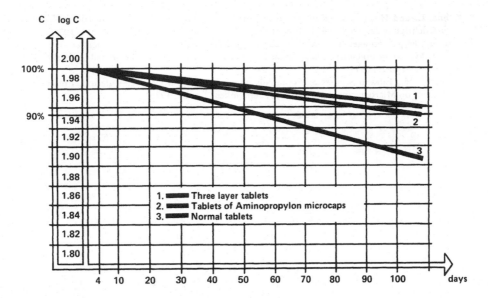

FIGURE 5. Aminopropylon microcaps — Tri oh lungo tablets — stability. Hydroxycobalamin at 40°C R.H. 75%.

ured on the most reactive (hence least stable) substance in the forum formulation, namely hydroxycobalamin. As can be seen in the graph, both the layered tablet and the tablet made with microencapsulated aminopropylon proved materially more stable than the ordinary tablet; on the other hand, the stability obtained with microencapsulation did not differ significantly from that provided by the more costly layering process.

These, then, are the outstanding applications of microencapsulation in the drug industry, at least as regards processes that modify the performance of the finished dosage form. There are a few others, which do not modify performance, but simply facilitate the material handling of active substances (for instance by improving the flowability of sticky powders) or provide worker protection from potentially noxious agents; these, however, are not included in this presentation.

REFERENCES

1. Saggers, V. H., Chasseaud, L. F., and Cooper, A. J., Plasma concentrations, duration of analgesia, and bio-availability of microencapsulated aspirin (Levius), *Clin. Trials J.*, 9, 36, 1972.
2. Simplicio, F., Scagliusi, P., and Sozio, N., Meccanismo di azione e indicazioni dell'aspirina coacervata, *Reumatismo*, 24, 183, 1972.
3. Vignalou, F. and Beck, H., Etude clinique et biologique de deux formes d'aspirine: estimation de la tolérance gastrique, *Thérapie*, 22, 967, 1967.
4. Lechat, P., Ganter, P., Fontagne, J., and Flouvat, B., Etude expérimentale de la tolérance gastrique d'une aspirine en micrograins enrobes, *Therapie*, 22, 403, 1967.
5. Maggi, G. C. and Coppi, G., Therapeutic activity and effects on fecal blood loss of a new microencapsulated potassium chloride preparation, *Curr. Ther. Res. Clin. Exp.*, 21 (5), 676, 1977.

6. **Kühns, L. and Hospes, K.**, Klinische Bedeutung und Anwendung eines Kaliumdefizit-Testes unter Berücksichtigung der Therapie mit Prednison, *Schweiz. Med. Wochenschr.*, 86, 783, 1956.

7. **Maccari, M.**, Avantages de la Microencapsulation des Médicaments, en Particulier en ce qui Concerne le Masquage du Goût de la Dicloxacilline et de la Flucloxacilline, presented at Ipharmex, Lyon, November 17 to 19, 1977.

8. **Calanchi, M.**, Eurand Studies on Some Advantages of Drug Microencapsulation, Presented at 3rd Int. Symp. on Microencapsulation, Tokyo, September 21 to 22, 1976.

Chapter 6

MICROENCAPSULATION BY INTERFACIAL POLYMERIZATION TECHNIQUES — AGRICULTURAL APPLICATIONS

Robert C. Koestler

TABLE OF CONTENTS

I. INTRODUCTION

When the relatively safe, but overly persistent, insecticides widely used in the 1960s fell into environmental and political disfavor and were phased out, they were often replaced by insecticides that had very short persistence in the field and much higher mammalian toxicity than the compositions they replaced. The use of microencapsulation to formulate agricultural products thus became attractive, as microencapsulation offered the possibility of overcoming these serious disadvantages of high reactivity and high toxicity. This chapter will describe a method of microencapsulation of agricultural products by interfacial polymerization, discuss possible reasons for their effectiveness, summarize the status of their commercial development, and give details of some of the advantages that have been observed.

II. MICROENCAPSULATION BY INTERFACIAL POLYMERIZATION

The technique of interfacial polymerization has been a favorite demonstration of college chemistry lecturers. In a beaker, a layer of diamine in water is carefully poured over a layer of diacid chloride (such as adipoyl chloride) in an organic solvent. A polymer, forming instantly at the interface, can be grabbed with a forceps and pulled from the beaker in the form of a nylon rope, with more polymer continuing to form at the interface as the rope is withdrawn.

Microencapsulation by interfacial polymerization can make use of this same reaction with some fundamental changes in the operating procedure. Instead of the reactants being layered, the organic phase containing the diacid chloride is dispersed in the aqueous phase before addition of the amines. Dissolved also in the organic phase is the active ingredient (pesticide) to be encapsulated. Formation and stabilization of the emulsion is accomplished by the addition of a low concentration of a nonionic surfactant and use of a high-shear stirring apparatus, producing a dispersion of oil droplets containing a diacid chloride throughout a continuous water phase. The oil droplets are very small and may range from one to several hundred microns in diameter. Both the particle size and the particle size range can be controlled. The addition of a diamine to the aqueous phase of the emulsion results in an almost instantaneous polymerization at the interface of each oil droplet to form a film completely enclosing the droplet. Beause the diacid chloride is not soluble in the aqueous phase and because the diamine is soluble in the organic phase, the polymer wall grows inward into the capsule. After the initial polymer film is formed, the diamine diffuses through the polymer in order to react with the diacid chloride. As the polymer wall gets thicker, the rate of diffusion of the diamine decreases. The polymerization reaction ceases when all of the diacid chloride has reacted either with the diamine to form polyamide or with water in a chain terminating hydrolysis to acid. Generally, an acid acceptor is used in the aqueous phase to remove the HCl formed and thereby keep all of the diamine available for polymer formation.

The microcapsules formed by this technique may be filtered off and dried, in some cases to a free-flowing powder containing up to 90% liquid payload of the active ingredient. For applications where a flowable, sprayable formulation is desired, the capsules may be allowed to remain as an aqueous slurry. In this case, a sieving step removes oversize capsules so that spray nozzles will not be clogged during use. For improved storage stability and ease of use, a thickening or suspending agent is added to prevent settling and compacting of the capsules. Materials that can be used as suspending agents include hydroxyethyl cellulose, gums, clays, submicron-size silica, and other inorganic materials.

TABLE 1

Polymer Systems For Microencapsulation by Interfacial Polycondensation

Oil phase monomer	Aqueous phase monomer	Polymer
Dicarboxylic acid chloride	Diamine	Polyamide
Dicarboxylic acid chloride	Diol or polyol	Polyester
Diisocyanate	Diamine	Polyurea
Diisocyanate	Diol or polyol	Polyurethane
Bis(chlorocarbonate)	Diol or polyol	Polycarbonate
Bis(sulfonyl chloride)	Diamine	Polysulfonamide

TABLE 2

Crosslinking Monomers

1,3,5-Benzenetriamine	1,3,5-Benzenetricarboxylic acid chloride
2,4,6-Triaminotoluene	1,2,4,5-Benzene tetracarboxylic acid chloride
1,3,6-Triaminonaphthalene	1,3,5-Benzene trischloroformate
Polymethylene polyphenylisocyanate	
Triphenylmethane-4,4′,4″-triisocyanate	

Active ingredients that are best encapsulated by the interfacial polymerization system are organic liquids (or low- melting solids) of very low water solubility that contain no functional groups that would react with the encapsulating system. Many of the most widely used pesticides fall in this class. Both operationally and economically, the encapsulation of such materials is attractive because the continuous phase and storage phase is water. The resulting products carry a high payload and can be formulated at concentrations of 2 or 3 lb of active ingredient per gallon.

Insoluble solids can be encapsulated as a slurry in an organic liquid carrier. However, active ingredient payloads are low, and the method of release is not easily understood.

The encapsulation of water-soluble materials is feasible by the method of interfacial polymerization, although there are problems that make the production of a cost-effective product difficult. Capsule payloads are reduced by the large amount of water they must carry. The organic continuous phase of some applications must be recovered and purified or freed from unreacted monomer because in the encapsulation of water soluble materials the diacid chloride is often used in excess.

Organophilic compounds of greater than 1% water solubility, primary and secondary amines, alcohols, mercaptans, and materials sensitive to either high or low pH are best encapsulated by methods other than interfacial polymerization.

III. PARAMETERS AFFECTING CAPSULE PROPERTIES

The interfacial polymerization reaction requires the condensation of complementary, organic polycondensate-forming intermediates. Polymers that may be produced by this technique include polyamides, polyureas, polyesters, polysulfonamides, and polycarbonates (Table 1).

Crosslinking of the polymer wall has been found desirable to achieve durable, storage-stable capsules.[i] A sufficient amount of a polyfunctional reactant (i.e., trifunctional or greater) is therefore included in the polycondensation recipe for this purpose (Table 2).

The wall thickness of the microcapsules is a function of the ratio of the amount of polymer formed to the amount of internal phase (encapsulate) and is also a function of the capsule size. Although a capsule having a thick wall might be more durable and hold its contents more securely, it is also less economical to produce and carries less payload. Conversely, a thin-walled capsule costs less to produce and carries a greater payload, but it may be more fragile and give less protection to its contents. Thus, capsule wall thickness will affect the efficacy and the toxicological properties of the product, as well as the cost of producing it.

Particle size is an important variable. It is held within a fairly narrow range because the capsules must not be so large as to clog the spraying nozzles or the filtering screens of agricultural spraying equipment. Conversely, very small capsules would have excessively thin walls and would break easily. The following relationship between capsule size and wall thickness is derived assuming the density of the encapsulate and the polymer wall are the same and equal to 1.0:

$$W = r_1 - r_2 \tag{1}$$

where w = wall thickness; r_1 = capsule radius; r_2 = radius of a droplet of encapsulate.

$$V_1 = \frac{4}{3}\pi r_1^3 \tag{2}$$

$$V_2 = \frac{4}{3}\pi r_2^3 \tag{3}$$

where V_1 = volume of capsule; V_2 = volume of droplet of encapsulate. Because the density of the polymer and the encapsulate are assumed to be the same and equal to 1.0, the volume of the polymer v_P can be calculated from its weight and

$$V_P = V_1 - V_2 \tag{4}$$

$$V_2 = V_1 \left(\frac{100 - \% \text{ wall}}{100} \right) \tag{5}$$

Then:

$$W = \sqrt[3]{\frac{3V_1}{4\pi}} - \sqrt[3]{\frac{100 - \% \text{ wall}}{100} \left(\frac{3V_1}{4\pi} \right)} \tag{6}$$

Figure 1 shows the calculated capsule wall thickness as a function of capsule diameter and volume. It can be seen that a 27-μm capsule with a 10% wall by weight would have a wall thickness of about 0.5μm. Each capsule would have a volume of about 10^{-8} ml, and it would take 10^8 capsules to contain 1 g of encapsulate. The capsule payload would be 90%.

IV. THE EFFECT OF MICROCAPSULE GEOMETRY ON RESIDUAL ACTIVITY AND EFFICIENCY

It has been suggested[2] that PENNCAP-M®* INSECTICIDE capsules, rather than releasing by a diffusion-controlled mechanism, release by leakage of the contents through numerous holes in the capsule wall. Thus, rather than the release being con-

---

* Encapsulated methyl parathion, a product of the Pennwalt Corporation, King of Prussia, Pa.19406.

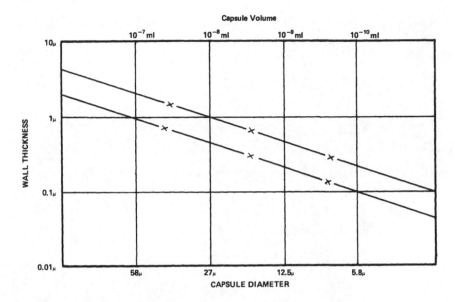

FIGURE 1.   Wall thickness vs. capsule volume and capsule diameter.

trolled by first-order kinetics, the spherical geometry of the system was said to be the controlling factor. That this theory could be at least partly correct is supported by scanning electron microscope photographs of microencapsulated xylene, showing the capsule surface to be a lattice rather than a solid membrane when viewed at magnification × 10,000. However, scanning electron microscope photos of encapsulated methyl parathion do not show this same type of capsule surface: at magnification × 10,000 there are no surface holes evident. The surface morphology of the capsules instead resembles a tennis ball, slightly fuzzy, but no visible holes (Figure 2). Scanning electron microscope photographs of capsules containing various other materials have different appearances, (Figures 3,4,5) suggesting that the morphology of a capsule depends primarily on what is inside it. This is not surprising because changing the encapsulate changes the characteristics of the interface between the two liquid phases where polymerization takes place. Capsule morphology may play a major role in the release characteristics of a capsule and in the interaction of a capsule with a substrate. A very rough capsule wall would have more surface area than a smooth capsule wall and might be expected to have more thin spots and imperfections, all of which would lead to faster release.

The well-defined spherical geometry of the system makes it possible to relate efficiency and persistence for an encapsulated insecticide.[3] The spraying of an aqueous emulsion on an area is generally an attempt at 100% coverage of the area with a very thin layer of the pesticide. On the other hand, the spraying of a microencapsulated formulation at the same active ingredient rate does not cover 100% of the area. Theoretical calculations show, for example, that for 27-μm microcapsules, only 0.6% of the area has contact with the pesticide (Figure 6).

Because microencapsulated formulations cover such a small percent of the sprayed area, questions may arise concerning the reasons for the effectiveness of this type of product. Several factors must be considered. First, standard aqueous emulsions are sprayed to get 100% area coverage, so local concentrations are very low. Therefore, the active ingredients are prone to absorption by the substrate and for certain chemical compositions, deactivation by ultraviolet light from the sun, oxygen, or moisture from the air or other degradation processes. Encapsulated active ingredients, however, occur

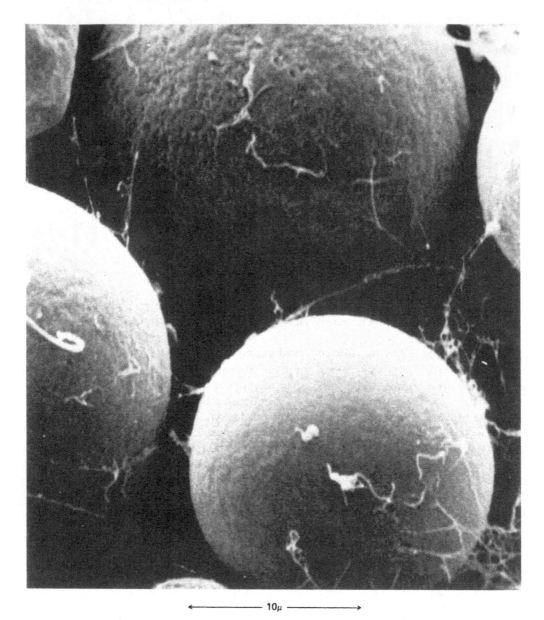

FIGURE 2.   Microencapsulated methyl parathion. (10 μm).

in areas of high local concentration, are more likely to saturate adsorbing sites, and may be more resistant to ultraviolet degradation and chemical reactions because of protection by the capsule walls and absorption by the thicker layer of pesticide itself.

The second factor that must be considered is what are the chances of an insect stepping on, bumping into, or eating a capsule? Putting it another way, how far does an insect have to walk between capsules? Calculations show that under idealized conditions, for 27-μm capsules, at a spray rate of 1 lb of active ingredient per acre, the center-to-center capsule spacing would be approximately 0.3 mm (Figure 7). Thus, even with an area coverage of less than 1%, the spacing between the capsules is very small, and it is doubtful that an insect walking across a sprayed leaf could escape contact.

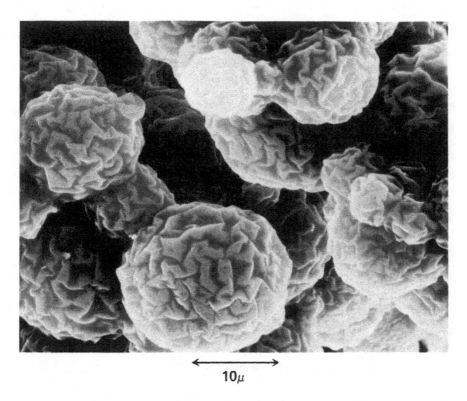

10μ

FIGURE 3.   Microencapsulated Sumithon®. (A product of Sumitomo Chemical Company, Osaka 541, Japan.) (10 μm; magnification × 2500.)

Capsules in the size range of 10 to 100 μm are generally most useful because they exhibit increased residual activity and high efficiency due to close capsule spacing. They have less tendency to drift than capsules under 10μm and are sprayable through conventional ground and air equipment. Capsules 10μm and larger settle rapidly when sprayed and therefore present less of an inhalation hazard than smaller capsules.

V. COMMERCIAL PRODUCTS

The first microencapsulated product sold in the U.S. for agricultural use was PENN-CAP-M® INSECTICIDE, encapsulated methyl parathion, manufactured by the Pennwalt Corporation.[4-10] Label registration was granted by the Environmental Protection Agency (EPA) in 1974 for commercial marketing.[11] The initial registration was for use on alfalfa, cotton, and sweet corn. By 1977, additional registrations had been granted to include use on a total of 16 major crops. Other crops are under investigation, and applications for registration and petitions for tolerance exemptions are being submitted to the EPA as additional data becomes available.

The capsule wall of PENNCAP-M® is a crosslinked polyamide-polyurea polymer. The capsules are formed by allowing sebacoyl chloride and polymethylenepolyphenylisocyanate to react with a mixture of ethylenediamine and diethylenetriamine at the interface of a microdroplet of methyl parathion suspended or dispersed in water. The EPA required data showing that the residues of the capsule wall material are inert and nontoxic and in 1974 granted the wall material an exemption from the requirement of tolerance on certain protected crops (Figure 8). The amount of testing that was required before the issuance of the commercial label and clearance of the wall material

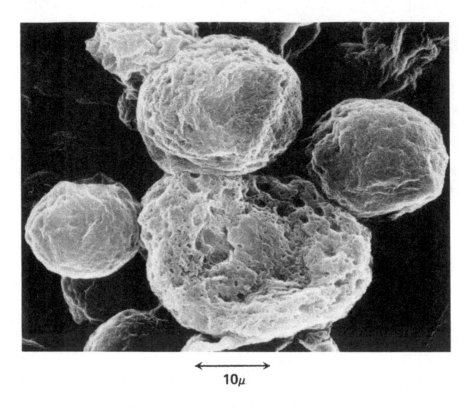

10μ

FIGURE 4.   Microencapsulated malathion. (10 μm; magnification × 2000.)

was similar to that required for a new pesticide. Tolerances for methyl parathion were established previously.

VI. DEVELOPMENTAL PRODUCTS

Microencapsulated ethyl parathion, PENNCAP-E® INSECTICIDE, is currently under development. Physically the produce is similar to PENNCAP-M® INSECTI-CIDE. An experimental permit for use against certain insects on sorghum has been granted by the EPA.

Another developmental product currently being evaluated under an experimental permit is microencapsulated Diazinon®* (Figure 9) called KNOX OUT® 2FM IN-SECTICIDE. An application for a commercial label to control household insects is currently pending. KNOX OUT® 2FM is a flowable, microencapsulated formulation containing 2 lb of Diazinon® per gallon and is physically similar to PENNCAP-M®. The experimental permit specifies use against domestic pests, such as cockroaches, ants, spiders, ticks, and numerous lawn insects. KNOX OUT® 2 FM has been found to be more effective and more persistent than emulsifiable concentrate formulations. The product is nontoxic at a dose of 21 g/kg (acute oral — rats), at least a 14-fold advantage over the emulsifiable concentrate, and nontoxic at a dose of 10 g/kg (acute dermal — rabbits), both based on formulation.

VII. EXPERIMENTAL PRODUCTS

Particularly suited for microencapsulation are pheromones, (Figure 10) highly ex-

* Diazinon®—Trademark of Ciba-Geigy Corporation, Ardsley, N.Y.10502.

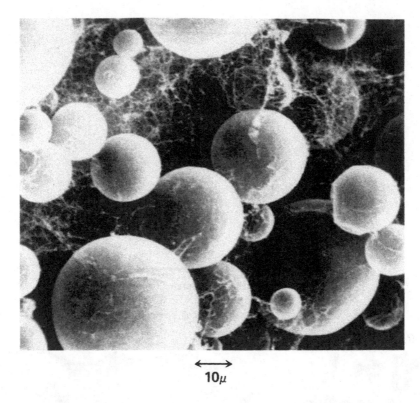

FIGURE 5.   Microencapsulated synthetic pyrethrum. (10 μm; magnification × 1000.)

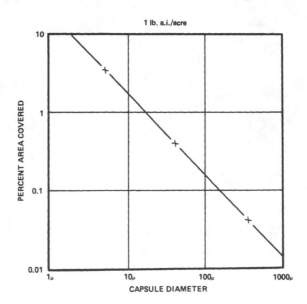

FIGURE 6.   Percent area coverage vs. capsule diameter.

pensive and volatile substances used in small quantities over vast acreage to disrupt the mating activities of insects. If, by a microencapsulated formulation, sufficient vapor concentration of a pheromone could be maintained in the air over a host crop for

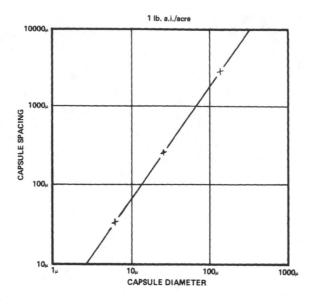

FIGURE 7.   Capsule diameter vs. capsule spacing.

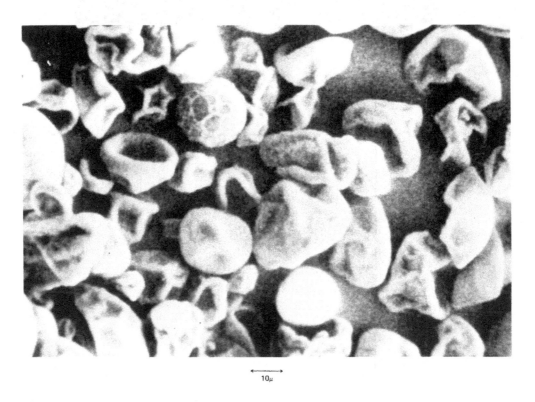

FIGURE 8.   Methyl parathion capsules after release of contents. (10 μm).

the entire mating season, it would prevent the insect population from building up, thereby minimizing economic damage to the crop by the insect.

Microencapsulated disparlure [sex attractant of the gypsy moth, *(Porthetria dispar*, L.)] has been used in small-scale field tests in which it remained effective for

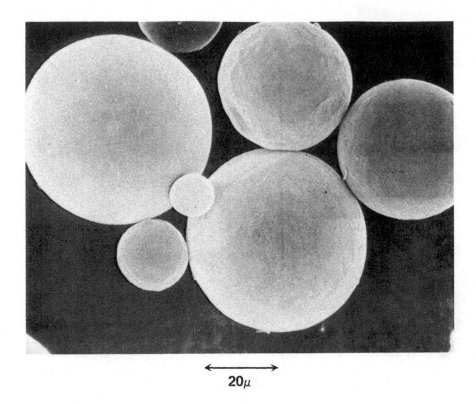

FIGURE 9.   Microencapsulated Diazinon®. (A product of Ciba-Geigy Corporation, Ardsley, N.Y. 10502) (20 μm; magnification × 1000.)

more than 6 weeks.[12] The U.S. Department of Agriculture has run large-scale field
tests that involved the spraying of encapsulated disparlure on gypsy moth infestations
of 1000 acres in Pennsylvania and 10,000 acres in Michigan for the purpose of disrupting mating between the male and female moths. Results on the large-scale tests were
disappointing, causing a return to the laboratory for additional research on the attractant itself.

Microencapsulated formulations of the pheromone of the redbanded leaf roller (*Ar-
gyrotaenia velutinana,* Wlk.) a serious pest of apples, have been field tested in orchards
in New York State. A 75 to 99% disruption of the attraction of the male redbanded
leaf roller to monitoring traps was observed.[13]

Mixtures of microencapsulated trimedlure, the sex attractant of the Mediterranean
fruit fly (*Ceratitis capitata,* Wiedemann), with microencapsulated insecticides are presently under investigation at the Hawaiian Fruit Fly Laboratory of the U.S. Department
of Agriculture. Residual effects of the attractant of up to 1 month in the field have
been observed. The spot-treatment technique is used. The pests are attracted to the
insecticide by the sex attractant, whereas the natural enemies remain unaffected.[14]

Microencapsulated formulations of (z)-7-dodecen-1-ol acetate (cabbage looper
pheromone) were tested by spraying on cabbage.[15] The capture of male cabbage loop-
ers in traps was reduced by 80% over a period of 24 days, implying that mating would
have been reduced by a like amount over that period.

Some of the problems associated with herbicides that could be modified by microencapsulation are lack of persistence, excessive volatility, phytotoxicity, and sensitivity
to ultraviolet light. Experimental microencapsulated chlorpropham formulations in
some field tests were herbicidally effective for a significantly longer period of time

20μ

FIGURE 10.   Microencapsulated insect pheromone. (20 μm; magnification × 450.)

than the commercial formulation.[16] In addition, the loss of active ingredient by vaporization was greatly decreased.[17]

VIII. TEST RESULTS — PENNCAP-M® INSECTICIDE

PENNCAP-M® has been found to give excellent results when used on alfalfa against weevils and aphids. It was the first insecticide specifically registered for control of the blue alfalfa aphid (*Acyrthosiphon kondoi*), a new pest in California and Arizona. In one California trial, PENNCAP-M® gave the highest kill and longest control time of all insecticides tested on blue alfalfa aphids.[18]

PENNCAP-M® controls pests of tobacco, such as tobacco budworms, hornworms, and aphids. In one tobacco research study, use of PENNCAP-M® allowed treatment on a 14-day schedule instead of the 10-day schedule required by the emulsifiable concentrate. Per acre yield was increased 30% over an untreated control.

PENNCAP-M® effectively controls Mexican bean beetle, bean leaf beetle, and other major soybean insect pests.

Used on apples and pears, PENNCAP-M® is a highly effective insecticide for controlling codling moth and in the Northwest, San Jose scale. Even on such sensitive apple varieties as McIntosh and Golden Delicious, PENNCAP-M® has caused no russetting. Field experience with PENNCAP-M® demonstrates a variety of advantages that are possible through microencapsulation.

A. Residual Activity

PENNCAP-M® at ¼ lb active ingredient per acre has been found to control some

TABLE 3

PENNCAP-M® On Sweet Corn

Treatment	Rate per acre (lb)	Spray schedule (days)	Damaged ears[a] (%)
PENNCAP-M®	0.5	1	7.34
PENNCAP-M®	1.0	1	0.66
PENNCAP-M®	0.5	2	6.00
PENNCAP-M®	1.0	2	4.66
Methyl parathion E.C.	0.5	1	3.34
Methyl parathion E.C.	1.5	2	16.0
Check	0	2	76.66

[a] Anything with 5% damage or less is considered U.S. No. 1 Fancy Grade.

TABLE 4

PENNCAP-M® on Cotton

Treatment	Rate per acre (lb)	Spray schedule (days)	Seed cotton per acre (lb)
PENNCAP-M®	1.0	5	2148
Methyl parathion E.C.	1.0	5	1613
PENNCAP-M®	0.5	5	1921
Methyl parathion E.C.	0.5	5	1016

TABLE 5

PENNCAP-M® vs. Japanese Beetles on Beans

Treatment	Pounds per acre	Percent killed after			
		2 days	3 days	5 days	8 days
PENNCAP-M®	0.5	100	100	100	80
Methyl parathion E.C.	0.5	100	100	0	0
Check	—	0	0	0	0

insects up to 5 to 7 days in the field. Under the same conditions, a standard E.C. formulation of methyl parathion gave control for only 1 to 2 days. The use of PENN-CAP-M® frequently requires fewer applications and results in less insecticide used per acre. In a test on sweet corn in Alabama, the results shown in Table 3 demonstrate the increased effectiveness of PENNCAP-M® at the 2-day spray schedule.[20]

Superior insect control can lead to increased yields as demonstrated by the test results shown in Table 4 for cottonseed yields in Georgia.[20] The results of a test on beans in Pennsylvania demonstrate the increased residual effect of PENNCAP-M® vs. the Japanese beetle, as shown in Table 5.[20]

B. Toxicity

Encapsulated methyl parathion has an oral toxicity at least five times less than the standard formulation. The acute oral toxicity (LD_{50}) for the standard emulsifiable concentrate is 10 to 20 mg/kg(mice) based on active ingredient, whereas for PENNCAP-M® it generally is more than 100 mg/kg. The dermal toxicity (LD_{50}) on an active ingredient basis of the emulsifiable concentrate is generally considered to be 100 mg/

kg (rabbits), whereas for PENNCAP-M®, it is greater than 1200 mg/kg. The apparent contradiction of increased insect toxicity and decreased mammalian toxicity may at first seem puzzling. As long as the hydrophilic capsules are dispersed in the aqueous phase, little of the hydrophobic methyl parathion appears at the capsule surface. As the sprayed capsules dry out, the methyl parathion comes to the surface available for insect contact. The result of this decrease in toxicity is that the insecticide is much safer to handle than methyl parathion emulsifiable concentrate, and no poison label is required by the Department of Transportation or the EPA.

C. Efficiency

An important advantage of microencapsulation can be its ability to require the use of less toxicant to do the job. For instance, in a test[20] on sweet corn, ¼ lb/acre was as effective as ½ or 1 lb/acre of the emulsifiable concentrate. Thus, we can make the supply of the insecticide go further and at the same time decrease environmental contamination.

D. Selectivity

In a test[20] on alfalfa in Utah, PENNCAP-M® at 8 oz/acre gave 92% weevil control and had little effect on beneficial insects. Furadan®* at 8 oz/acre gave 100% control, but also killed virtually all beneficial insects.

E. Phytotoxity

The phytotoxicity of methyl parathion represents a major obstacle to its use on apples. PENNCAP-M®, however, when used at recommended rates, has exhibited no phytotoxicity to apples and has been found to be an excellent product for insect control on this crop.[19]

F. Volatilization Loss

Methyl parathion vapor has been measured adjacent to fields sprayed with either encapsulated or standard methyl parathion formulations. The measurements indicated that encapsulation reduced the methyl parathion concentration in the air adjacent to the treated fields by as much as a factor of ten.[21]

G. Problems

As can be expected with any new product, problems may arise from time to time as the use increases and the applications become more varied with expanded label coverage.

The most serious problem to beset PENNCAP-M® thus far is its toxicity to bees. Methyl parathion itself is very toxic to bees. Methyl parathion or PENNCAP-M® should never be used on crops or weeds in bloom where bees are actively foraging. Failure to follow this directive which appears on the PENNCAP-M® label can be devastating to bees. When pollen contaminated with any beetoxic insecticide, including PENNCAP-M®, is carried back to the hive, bees and larvae may be killed by eating the pollen. Following an initial flurry of bee-kill reports allegedly involving PENN-CAP-M®, careful investigation of the cases generally indicated applications inconsistent with the present label or pointed to other products. As the situation stands now, with users following the labeling of the product and aware of its hazards to bees, few bee-kill reports were received during the 1977 growing season.

*   Furadan®—Trademark of FMC Corporation, Middleport, N.Y.14105.

IX. ADVANTAGES OF MICROENCAPSULATION

The removal of DDT and other persistent chlorinated materials from the pesticide roster in the early seventies forced farmers to replace these materials of relatively low human toxicity with products of short persistence, but high mammalian toxicity. It is these two serious disadvantages that microencapsulation originally sought to correct. As more experience has been gained in the use of microencapsulated pesticides, advantages in addition to the increased residual activity and reduced mammalian toxicity that have been achieved have become apparent. These include superior insect control, increased crop yields, fewer pesticide applications, less active ingredient used per acre, higher efficiency, reduced phytotoxicity, increased toxicity to resistant species, and reduction of pesticide loss due to volatilization. Microencapsulation of pesticides by interfacial polymerization is a relatively new technique that has demonstrated significant advantages and shows great promise for increasing use in the agricultural field.

REFERENCES

1. **Lowell, J. R., Jr., Culver, W. H., and DeSavigny, C. B.,** Effects of wall parameters on the release of active ingredients from microencapsulated insecticide, in *Controlled Release Pesticides,* ACS Symposium Series No. 53, Scher, H. B., Ed., American Chemical Society, Washington, D. C., 1977.
2. **Collins, R. L.,** Microgranules and micro-encapsulation of pesticides, in *Proc. Int. Controlled Release Pesticide Symp.,* Harris, F. W., Ed., Wright State University, Dayton, Ohio, 1975, 105.
3. **Koestler, R. C.,** A theory of a mechanism of action of encapsulated herbicides and insecticides, in *Proc. Int. Controlled Release Pesticide Symp.,* Cardarelli, N. F., Ed., University of Akron, Ohio, 1976, 8.1.
4. **Ivy, E. E.,** PENNCAP-M®: an improved methyl parathion formulation, *J. Econ. Entomol., 65,* 473, 1972.
5. **Vandegaer, J. E.,** U.S. Patent 3,577,515, 1971.
6. **Ruus, H.,** U.S. Patent 3,429,827, 1969.
7. **Vandegaer, J. E.,** *Microencapsulation: Processes and Applications,* Plenum Press, New York, 1974.
8. **Vandegaer, J. E. and Meier, E. G.,** U.S. Patent 3,575,882, 1971.
9. **Santo, J. E.,** U.S. Patent 3,667,776, 1971.
10. **DeSavigny, C. B.,** U.S. Patent 3,959,464, 1976.
11. **Anon.,** Microencapsulated pesticide reaches market, *Chem. Eng. News,* July 29, 15, 1974.
12. **Beroza, M., Stevens, L. J., Bierl, B. A., Phillips, F. M., and Tardif, J. G. R.,** Pre and postseason field tests with Disparlure, the sex pheromone of the gypsy moth, *Environ. Entomol., 2,* 1051, 1973.
13. **Carde, R. T., Trammel, K., and Roelofs, W. L.,** Disruption of sex attraction of the redbanded leafroller with microencapsulated pheromone components, *Environ. Entomol., 4,* 448, 1975.
14. **Keiser, I., Kobayashi, R. M., and Harris, E. J.,** Enhanced duration of residual effectiveness against the Mediterranean fruit fly of Guava foliage treated with encapsulated insecticides and lures, in *Proc. Int. Controlled Release Pesticide Symp.,* Harris, F. W., Ed., Wright State University, Dayton, Ohio, 1975, 264.
15. **McLaughlin, J. R., Mitchell, E. R., and Tumlinson, J. H.,** Evaluation of some formulations for dispensing insect pheromones in field and orchard crops, in *Proc. Int. Controlled Release Pesticide Symp.,* Harris, F. W., Ed., Wright State University, Dayton, Ohio, 1975, 209.
16. **Gentner, W. A. and Danielson, L. L.,** The influence of microencapsulation on the herbicidal performance of chloropropham, in *Proc. Int. Controlled Release Pesticide Symp.,* Cardarelli, N. F., Ed., University of Akron, Ohio, 1976, 7.26.
17. **Turner, B. C., Glotfelty, D. E., Taylor, A. W., and Watson, D. R.,** Volatilization of Microencapsulated and Conventionally Applied CIPC in the Field, Paper 8, Pesticide Division, in Abstracts of Papers, 173rd American Chemical Society National Meeting, New Orleans, March 20 to 25, 1977.

18. Blue aphid threatens alfalfa, *Agrichem. Age,* 19 (3), 6, 1976.
19. **Lowell, J. R., Jr. and Murnigham, J. J.,** Economics of Microencapsulation of Pesticides, Paper 8, Chemical Marketing and Economics Division, in Abstracts of Papers 172nd American Chemical Society National Meeting, San Francisco, August 29 to September 3, 1976.
20. Unpublished data, Agchem Division of Pennwalt Corporation, Fresno, Calif.
21. **Jackson, M. D. and Lewis, R. G.,** Effect of Physical Form on Volatilization of Methyl Parathion from Treated Fields, Paper 6, Pesticide Division, in Abstracts of Papers, 172nd American Chemical Society National Meeting, San Francisco, August 29 to September 3, 1976.

Chapter 7

THE WURSTER PROCESS

Harlan S. Hall and Ralph E. Pondell

TABLE OF CONTENTS

I. THE PROCESS

The Wurster process is a technique for applying coatings around particles. The process was invented in 1959 by Professor Dale E. Wurster,[1-7] then of the University of Wisconsin School of Pharmacy. The patents were assigned to the Wisconsin Alumni Research Foundation which has administered the patents since that time.

In the Wurster process, the particles to be coated are fluidized on an upward-moving airstream. A high-velocity air stream is introduced into the fluidized bed, causing a spout. A cylindrical partition is placed around the spout formed by the high-velocity air stream to prevent the particles in the spout from colliding with the particles descending in the fluidized bed. In this manner, a cyclic flow of the particles is caused.

When the particles enter the high-velocity spout, they are accelerated and physically separated from each other. It is important to note that the flow at this point is rapid, but smooth. The particles are uniformly accelerated on an air stream, not harshly battered by a turbulent flow pattern.

As the high-velocity air and the particles move up the partition, the coating is applied by a spray nozzle mounted at the base of the spout. The coating applied can be selected from a very wide variety of materials. This selection will be discussed later.

The process air which moves the particles also serves to dry the coating. Because of the relatively large amounts of air used, excellent drying conditions are achieved in this process. By the time the particles clear the top of the partition, they are already dry to the touch. This very rapid drying is particularly useful when the solvent for the coating is also a solvent for the material being coated and when aqueous systems are used.

Once the air stream and particles clear the top of the partition, the air in the spout spreads out to fill the expansion chamber. As the air spreads out and slows down, the particles settle out onto the top of the bed of fluidized particles. Because the bed of particles is fluidized by air, additional drying occurs as the particles descend to the bottom of the bed and reenter the partition to again be accelerated by the high-velocity air stream and receive additional coating.

Particles cycling in this manner pass the nozzle every 6 to 10 sec, receiving additional coating with each pass. The process is allowed to continue until the desired amount of coating has been applied.

The particles to be coated may range in size from a few microns to large tablets or pieces. They may be spheres, ovals, cubes, discs, noodles, or simply irregular.

The equipment used is designed to provide conditioned process air for moving the particles. Atmospheric air passes through an inlet air filter, a heat source, into the air distribution chamber, and into the coating chamber. The air leaving the coating chamber enters the expansion chamber and is discharged from the system by a blower.

The air flow through the system is controlled by one or more dampers which are adjusted from the control panel. The air flow is monitored with a readout on the control panel.

The temperature and, if desired, the humidity of the process air is also controlled from the panel.

The air distribution chamber provides a baffle through which the process air is distributed uniformly to the base of the coating chamber.

The coating chamber consists of a perforated air distribution plate, a partition, and the atomizing nozzle. Above the coating chamber is an expansion chamber of greater diameter than the coating chamber in which the linear velocity of the air stream is decreased, allowing the particles to settle out. A filter can be fitted into the expansion chamber to prevent very fine particles from exiting the system on the air stream.

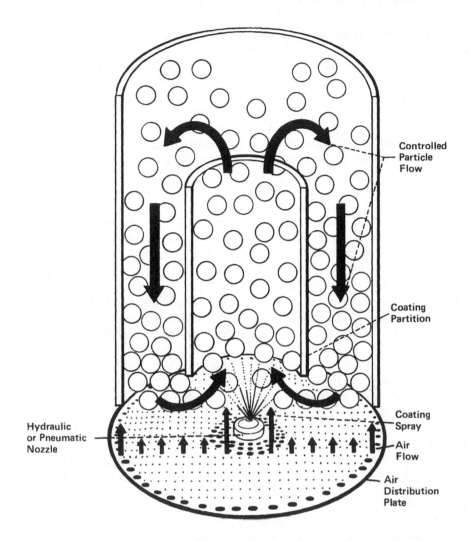

FIGURE 1.   Diagram of a Wurster coating chamber.

In preparing to coat in a Wurster unit, there are several conditions which must be determined in order to do an efficient job. Most units sold are fitted with a single replaceable air distribution plate and partition. Units which will be used for development work usually have three or more plates and may have three or four partitions. Having the correct plate design is critical to the successful coating of particles.

Selection of the proper partition, in combination with the proper plate, will result in a smooth flow of product in the coating chamber. The partition is adjustable to permit the operator to set the gap between the plate and the bottom of the partition as necessary. Fine powders, for example, less than 60 mesh (250 μm), require that the gap be relatively small, perhaps ¼ in. Larger particles, such as tablets, may require the gap to be 4 in. or more.

In loading the unit, the final volume of the material after coating must be considered, since the bed volume cannot exceed the height of the top of the partition at any time during the run.

Variables which the operator can adjust include the process air volume and temperature, the solution addition rate, the atomizing conditions, and the composition of the coating.

When the necessary amount of coating has been applied, the unit is shut down and the coating chamber removed for unloading. In most production situations, the unit is supplied with a second coating chamber which is loaded while the first chamber is in the unit. As soon as one chamber is removed from the unit for unloading, the second is moved into the unit and the coating cycle begun. While this chamber is operating, the first can be unloaded and reloaded. Alternatively, the unit can be designed with a single chamber and a dump assembly so that it is not necessary to remove the chamber for unloading.

A. Release Rate

Primarily, the rate of release for a product coated by this process depends upon the nature of the coating material. Examples include coatings which release the active by a variety of controlled mechanisms, including slow release, fast release, delayed release, release due to change in temperature, pH, or moisture, and release due to biological or chemical degradation of the film.

The release of an active is also a function of coating thickness. If a film is insoluble in a particular system, but permeable, the active will dialize out of the intact capsule. A detailed discussion of such membrane effects are found in Chapters 4 and 5 of Volume 1.

If the coating is partially compatible with the active core or if the active is actually part of an insoluble film, the active will migrate to the surface at a predictable rate. A discussion of migration of an active from or through a matrix is found in Chapters 1, 2, and 3 of Volume 1.

The coating may also be formulated to release the active ingredient as the coating is dissolved or eroded. Chapter 1 is a discussion of such erodible matrices.

In addition to the above, coatings may be sensitive to pH; thus, a coating can be formulated to be resistant to neutral water, but to dissolve in an acid or if you prefer, a base. Coatings can also be soluble in water, but resistant to an acid (or base).

Many potential coating materials are sensitive to temperature. Materials can be selected to melt from room temperature to several hundred degrees. This obviously permits the design of coatings for a variety of applications.

Fats, proteins, and starches are susceptible to attack by enzymes, providing yet another mechanism for release of an active.

A partial list of materials used successfully in the Wurster process can be found in Appendix 1 of this chapter.

By designing a coating to meet the intended use, it is possible to tailor the release to meet specific goals. The graph in Figure 2 indicates the range of release rates which can be achieved simply by altering the ratio of two ingredients in the coating.

B. Particle Size

In the Wurster process, one is essentially covering the surface of particles with a coating, that is, encapsulating. Because smaller particles have more surface area per unit weight, small particles require more coating to achieve the same level of protection.

This is demonstrated by comparing the amount of coating required to apply a coating 0.10 mm thick onto particles of various sizes. For purposes of comparison, Table 1 assumes a particle density of 1.3 g/cm³ and a density of 1.0 g/cm³ for the coating material.

From the above, it is clear that as the particle size decreases, more and more coating

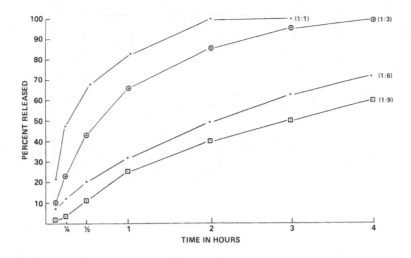

FIGURE 2. Release rates as a function of ingredient ratio.

TABLE 1

A Comparison of the Amount of Coating Required to Apply a Coating 0.10 mm Thick onto Particles of Various Sizes

	Uncoated particles			Coated particles		
U.S. mesh	Diameter (mm)	Particles per gram	Surface area per gram (mm²)	Coated diameter (mm)	Coating added (%)	Coating in prod. (%)
5	4.00	23	1,157	4.02	1.2	1.18
10	2.00	183	2,312	2.02	2.4	2.34
18	1.00	1,468	4,610	1.02	4.7	4.49
35	0.500	11,764	9,235	0.520	9.6	8.75
60	0.250	94,340	18,490	0.270	20.0	16.7
120	0.125	751,880	36,917	0.145	43.3	30.2
200	0.074	3,663,000	63,004	0.094	82.3	45.1
325	0.044	17,543,860	107,018	0.064	163.5	62.0

is required to obtain a given degree of protection. If a membrane of 0.10 mm is required to release an active at the desired rate, approximately 17.5 times as much coating or membrane will be required if the product is a 200-mesh powder than if it is an 18-mesh granule. If the 200-mesh powder is first formed into larger particles or allowed to agglomerate in the coating process, a significant economy in coating material can be achieved compared to coating the fine powder as discrete particles because the surface area will be greatly reduced.

A limiting factor in encapsulating smaller particles is the size of the atomized droplets of coating solution. When the droplets are large relative to the size of the particles, the particles are grossly wetted by the droplet and tend to stick together, i.e., agglomerate. Poor atomization, that is larger droplets, results in slower drying of the solvent due to the reduced surface area of the droplets. This slower drying further enhances the agglomeration of particles.

Figure 3 gives typical droplet sizes for commercially available nozzles. As indicated, air atomizing nozzles typically produce a finer spray than do hydraulic nozzles. For this reason, air atomizing nozzles are normally used when small particles are being

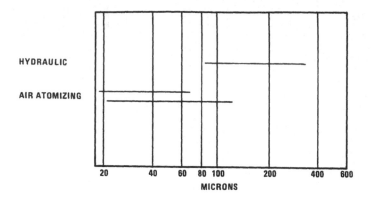

FIGURE 3.   Typical droplet sizes for commercially available nozzles.

TABLE 2

A Comparison of the Physical Properties of
Droplets of Various Sizes

Comparative subject	Particle size range (μm)
Smoke	0.001—0.01
Fumes	0.01—1.0
Dry fog	2.0—10
Wet fog	10—50
Misty rain	50—100
Light rain	100—500

coated. The size of the droplet formed will vary with the fluid being atomized, nozzle design, pressures used, and the rate at which the fluid is metered to the nozzle.

Table 2 compares the physical properties of droplets of various sizes.

In practice, the Wurster process is normally used when the desired particle size of the coated product is larger than 140 mesh (106 μm), although smaller coated particles have been prepared.

C. Solvents

For any solvent-based coating, a rate-limiting factor for any coating method is the drying of the coating, that is, evaporation of the solvent. The use of relatively large amounts of air in the Wurster process results in the rapid removal of solvent from the applied film. This can be particularly useful when aqueous coating systems are used. Regardless of whether the coating is a water solution, suspension, or emulsion, the important point is the rapid removal of water, especially if the product being coated is sensitive to water. It has been shown that it is possible to apply aqueous coatings onto water-sensitive substrates with little or no effect if care is used.[8]

The ability to remove water is a function of the humidity of the process air, the temperature of the process air, and the volume of air used. For aqueous coatings, this information is summarized conveniently in psychrometric charts. In Figure 4, ambient air at 70°F and 60% relative humidity (Point 1) is drawn into the unit and heated to 130°F, causing the relative humidity to drop to less than 10% (Point 2). The air then passes through the coating chamber, evaporating water and dropping in temperature to 93°F. At this point (Point 3), the water evaporated has increased the humidity of

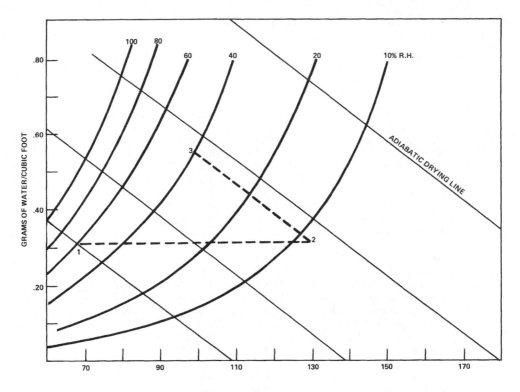

FIGURE 4. Psychrometric chart.

the air to 40% at 93°F. The air now contains 0.55 g of water per cubic foot compared to 0.30g /ft³ at ambient conditions. This difference of 0.25 g/ft³ means that using these conditions one can evaporate 250 ml of water per minute for each 1000 CFM of air used. In this manner, psychrometric charts can be used to determine the drying capacity or predict the temperatures needed for any conditions selected.

The same phenomena are observed when organic solvents are used. Normally, there is no solvent in the ambient air, but the evaporative cooling until saturation is reached still occurs. The situation for organic solvents can become very complex as one gets into mixed solvent systems. A good treatment of evaporation of mixed solvents from films can be found in the *Journal of Paint Technology*.[9]

The proper selection of solvent is a very important part of any coating system. To take best advantage of the good drying conditions in the Wurster process, one is usually seeking fast-drying solvents. Of course, one must still use solvents which will dissolve the film-forming material. The selection of solvents and solvent blends can be difficult, but the use of solubility parameters and evaporation rates can simplify this task. Crowley and associates[10] have combined the use of solubility parameters and hydrogen bonding and dipole moment to develop solubility maps for various resins. Hercules has published such maps for certain of their resins.[31]

In seeking fast drying solvents, one must take care to avoid being misled by boiling points. As Table 3 illustrates, boiling point is not a reliable indicator of evaporation rate. Most suppliers of solvents are able to provide charts of solvent properties, including relative evaporation rates, hydrogen bonding values, and solubility parameters. One should always evaluate the solvents with the film former in question, since the interaction of the film former and the solvent will often change the relative evaporation rates of solvents.

TABLE 3

Properties of Solvents

	Evaporation rate[a]	Distillation range (°F)	Solubility parameter
Methylene chloride	14.5	104-106	9.7
Acetone	7.7	132-134	10.0
Methyl acetone	7.0	122-153	9.5
Tetrahydrofuran	6.3	149-153	9.9
Carbontetrachloride	6.0	170-172	8.6
Methyl ethyl ketone	4.6	174-176	9.3
Ethylene dichloride	4.5	179-186	9.8
Trichloroethylene	4.5	188-190	9.3
Hexane	3.9	152-157	7.2
Heptane	3.9	201-207	7.4
Methyl alcohol	3.5	147-149	14.5
Ethyl alcohol, anhyd	1.9	165-176	12.7
Ethyl alcohol, 95%	1.7	165-176	12.7
Isopropyl alcohol, anhyd	1.7	179-181	11.5
Isopropyl alcohol, 91%	1.6	175-176	10.0
Methyl isobutyl ketone	1.6	237-241	8.4
Toluene	1.5	230-232	8.9
2-Nitropropane	1.1	242-252	10.7
Xylene	0.75	281-284	8.5
Iso-Butyl alcohol	0.63	223-229	10.5

[a]   *n*-Butyl acetate = 1.0.

In certain cases, one may be seeking other than the most rapid drying solvent because of a particular advantage of a different solvent. Such advantages include cost, safety in handling, safety of residues in product, solvent power for the resin, and compatibility with other ingredients or components of the product.

Table 3 is a partial list of solvents, their relative evaporation rates, distillation ranges, and their solubility parameters.

D. Temperature

Often the material you wish to encapsulate or the coating material or both will be heat sensitive. Experience has shown that it is possible to use process air temperatures in excess of the supposed temperature limit in some cases. This is so because of the evaporative cooling mentioned above. Figure 5 shows the temperature as measured at various points in an operating Wurster coating chamber.

In this illustration, the incoming process air is heated to 170°F. As the air and the particles pass the nozzle, the evaporation of the solvent cools the air. The temperature of the air exiting the top of the partition has dropped to 132°F. Above the partition, the 132°F air is mixed with the warmer air (138°F), fluidizing the descending particles, and the exit air is found to be 134°F. As the particles descend in the fluidized bed, they are moving countercurrent to the fluidizing air which is warmer deeper in the bed. At the base of the fluidized bed, the air is found to be 149°F. From this point, the particles are swept into the high-velocity air stream which is hotter, but at this point, the evaporative cooling of the solvent spray again cools the air stream to 132°F. Thus, in this example, it would be possible to coat a particle which cannot tolerate temperatures above 150° or 155°F, using air at 170°F.

This is illustrated in Figure 6 which shows the temperature profile for a complete

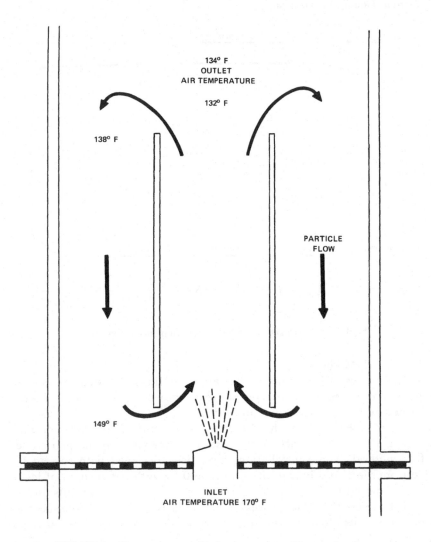

FIGURE 5. Temperature profile for an operating Wurster chamber.

coating run. In this case, the temperature of the processing air was raised to and held at 128°F. As the processing air began to heat up, the air exiting the unit also began to rise in temperature. After approximately 5 min, the coating spray was started, and the evaporative cooling caused the temperature of the outlet air to drop even though the temperature of the inlet air was still rising. The temperature of the air exiting the system equilibrated at about 70°F and remained at that temperature until the coating was complete and the spray turned off. When the spray was shut off, the temperature of the exiting air began to rise, and the unit was shut off and the coated product removed.

II. APPLICATIONS

Because the Wurster process can apply coatings from aqueous and organic solvents as well as latexes, melts, and suspensions, it has found application in many different fields. A large number of these are proprietary and cannot be discussed here; however, those that are covered will serve to demonstrate the versatility of this process.

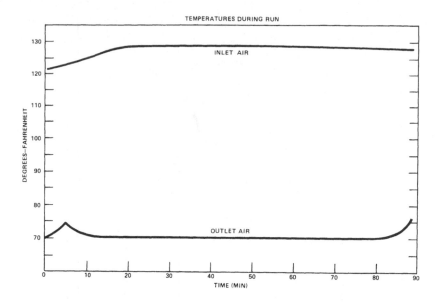

FIGURE 6.   Temperature profile for a complete coating run.

A. Pharmaceutical, Medicinal

The Wurster process was first invented to apply coatings onto pharmaceutical tablets and still does that extremely well. Over the years, however, the process has been continually refined and now is used for many purposes.

In the area of tablet coating, the Wurster process is used to produce a number of products. Several large pharmaceutical houses produce virtually all their coated products by this method. These installations have been designed with solvent recovery, complete automation, and in capacities up to 400 kg with batch times of less than 40 min.[11] Tablet coating is done commercially with both organic solvents and water as the vehicle for the coating. The process is particularly well suited to aqueous coating systems because of its ability to dry the water-based films before the water adversely affects the core.[8]

In addition to its applications for tablet coating, the Wurster process is also used to produce encapsulated actives in smaller sizes. Some of these products are encapsulated to prevent the patient from tasting the active. This is particularly important for pediatric drugs and drugs which must be taken frequently for an extended period of time. Of course, such products must be designed to release in the GI tract of the patient.

Other applications are found when it is desired to release the active over a period of time or at a specific point in the system. Examples of such products include enteric coated drugs which will not release in the stomach, but will release in the intestines[12,13] and also products which are designed to release at a uniform rate for a specific time.[14-16]

Pharmaceuticals are also encapsulated by this process to improve the stability of the active. Many biologically active products are unstable under particular conditions and would not have a practical shelf life if not protected. This is particularly true when the active is to be incorporated into a formulation with several other ingredients. Depending on what the active is sensitive to, the film applied may be an oxygen barrier, a water barrier, or simply physically segregate two or more reactive ingredients in the dry state.

In a similar vein, the Wurster process has been used to prepare oral vaccines. Dr.

John Debbie of New York State Department of Health, Albany has prepared an oral rabies vaccine for wild animals using this process.[17] One of the big problems was loss of activity when exposed to the conditions in the mouth and stomach. Dr. Debbie enteric-coated the vaccine and mixed it into a bait formulation. The animals ate the bait and were successfully immunized against rabies. Tests to date have shown the vaccine is effective on dogs, skunks, foxes, raccoons, and mongooses.

The process has also been used to coat adsorbents for use in blood purification. The reason such materials are coated is to make them selective. Adsorbents will remove a great many chemicals from the blood, many of them beneficial. By applying a semipermeable membrane around the adsorbents, they can be made more selective.[18,19] Such products can be used in hemodialysis for kidney failure, drug overdose, or poisoning.

B. Foods and Feeds

In the food industry, the Wurster process has been used to encapsulate vitamins and minerals, functional food ingredients, and also to reduce or eliminate caking of hygroscopic materials. Vitamins and minerals are frequently encapsulated to mask their characteristic flavors. In such applications, it is important that the particle size of the coated product be small enough to avoid contributing a "gritty" texture in the final food. It is clearly necessary that the nutrient be released in the GI tract or the purpose of adding it has been defeated. Taste masking requires virtually complete encapsulation, since a small amount of an undesirable flavor can ruin an otherwise fine product.

As with medicinal applications, many food ingredients are encapsulated to improve their stability. The stabilizing of vitamins is fairly commonplace, but other aplications include the protection of leavening agents from other ingredients to prevent premature reaction. Low-melting fats are often used for this application, since they will protect the active at room temperature or below, but will release quickly in a bake cycle. In certain cases, the coating may be water soluble, protecting the active only until the mixture is combined with water or milk.

Thin or partial coatings have proven effective in reducing the caking of hygroscopic materials. The presence of a layer of nonhygroscopic material between two hygroscopic particles will often prevent them from caking together even if they do pick up water.

All of the above apply to animal feeds as well as to foods for human consumption. In some cases, the benefit in animal applications will be greater than in humans, since no amount of logic will convince an animal to eat something distasteful.

C. Agriculture

In addition to simplifying the administration of vitamins and pharmaceuticals to livestock, the Wurster process has application in other agricultural areas.

In certain cases, the performance of pesticides can be enhanced by the development of controlled release formulations. The Wurster process has been used in a number of such products. An example of a somewhat unique product utilizing the concept of sustained release is cotton seed which has been coated with a polymer containing Di-Syston®. Prior to developing this product, cotton seed was treated by applying the insecticide directly onto the seed. This resulted in significant phytotoxicity to the seed and reduced germination and vigor. With untreated seed, germination and vigor were good, but the young seedling was very susceptible to insect predation. Moreover, it is necessary to apply the insecticide several times during the growing season, each application requiring manpower, equipment, and materials. By formulating the insecticide into the polymer and applying the polymer composition onto the seed, a number of

TABLE 4

Pesticide Performance (%) from Encapsulated Cotton Seeds

	Age of Plants				
Treatment	2 weeks	3 weeks	4 weeks	5 weeks	5 weeks, 3 days
Resin A	100%	100%	100%	87%	77%
Resin B	100%	100%	100%	95%	77%
Resin C	100%	100%	100%	100%	77%
Control Unencap- sulated	100%	100%	100%	37%	0%

TABLE 5

Greenhouse Data

Ratio of 50% Benomyl® to seed	Fungicide (lb/acre)	Bulbs diseased (%)
Control	None	90
2:1	0.4	8
4:1	0.8	3
6:1	1.2	0
8:1	1.6	0

the above problems are avoided and a few other benefits are gained.[20,21] First, the treated seed is more free flowing and easier to handle because the linters found on cotton seed, even delinted seed, are covered by the polymer and no longer cause the seeds to adhere to each other. Second, the seed is safer to handle because the pesticide is held in the polymer and will not dust off on the handler. Third, the insecticide is no longer phytotoxic to the seed, since it is not directly on the seed, but tied up in the polymer. Fourth, because the insecticide is released at a fixed rate by the polymer, it lasts longer than the same amount of insecticide applied by conventional methods (Table 4).

The idea of treating the seed rather than the entire field has also been applied to onions.[22] In this instance, onion seeds were coated with fungicide to protect the crop from white rot *(Sclerotium cepivorum),* Table 5 shows the level of control achieved by various levels of fungicide on the seeds. It is noteworthy that even the highest level of seed treatment, 8 parts fungicide (50% active) to 1 part seed by weight, uses only 1.6 lb of fungicide per acre compared to 10 to 15 lb per acre normally used to control this disease.

The encapsulation of Warfarin® in the Wurster process made it the first rodenticide to achieve Environmental Protection Agency (EPA) registration.[20-24] The rodenticide was encapsulated with a coating selected to cover the taste and odor of the active ingredient, yet release in the gut of the target animal. The coating prevents the rodent from detecting the presence of the toxin, thus it will eat more of the toxic bait and the effectiveness of the toxin is greatly improved. Figure 7 shows the improved palatability of the bait after the Warfarin® has been encapsulated.

A somewhat different approach was taken to achieve the desired release of a fish toxin, Fintrol®.[20,25,26] Fintrol® is a controlled release fish toxicant which is designed to release the active ingredient uniformly into a predetermined depth of water from

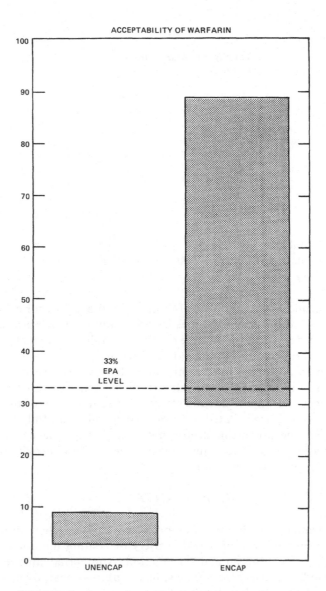

FIGURE 7. Improved palatability of encapsulated War-
farin®.

the surface. Fintrol® is a high-density inert core coated with a water soluble resin
which contains the toxin. The resin begins to dissolve immediately upon contact with
water and releases the toxin as it settles. The product is made in several formulations,
each designed to release toxin from the surface down to a predetermined depth. This
is illustrated by Table 6 which shows the toxicity of water treated with Fintrol®. No-
tice that once the water depth exceeds the design depth of the product, the product
stops releasing toxin. This is so because the toxin is released completely, leaving only
the inert core to continue to the bottom.

Seeds have also been coated to increase their size to permit machine planting.[27] By
increasing the size of certain very small seeds for machine planting, not only is the
waste of seed greatly reduced, but also the cost of thinning the excess plants is eliminated. In such cases, the primary objective is to increase the bulk of the seed, however,

TABLE 6

Toxicity of Water Treated with
Fintrol® 5

Test depth (ft)	Trout mortality (24 hr)
1	16/16
2	16/16
3	16/16
4	16/16
5	16/16
6	12/16
7	0/16

it is frequently desirable to include pH buffers, fungicides, trace nutrients, or other beneficial constituents in with the low-cost bulking agents.

D. Chemicals

A variety of chemicals which do not fall into the above categories have also been encapsulated by this technique. Examples include catalysts which are released to activate a system at the desired time. In these cases, the coating prevents the reaction from taking place before certain prescribed conditions are met.

Salts have been encapsulated to scavenge water from a reactive system which is inactivated by water. The addition of the salt directly has other undesirable effects which are avoided by encapsulating the salt and effectively keeping it out of the system.

Metallic particles have been coated to form particles with carefully defined electrical properties for use as the carriers in electrophotographic copy machines.[28-30] Such particles develop a static charge which causes the pigment to cling, providing a mechanism to transfer the pigment to the copy paper.

III. ADVANTAGES

A primary advantage of the Wurster process is the excellent drying conditions achieved. This permits the use of a wide variety of coating materials which permits great flexibility in product design and formulation. Another advantage of the process is its ability to handle a variety of particles, including spherical, oval, needlelike, and irregular, and may be high or low density. The particles may even be quite hygroscopic. The above, combined with the freedom to choose water- or solvent-based coatings or melts, means that seemingly endless combinations of materials can be combined to create new and unique products.

IV. DISADVANTAGES

The greatest disadvantage of this process is its inability to form extremely small capsules, that is, smaller than 75 μm. While particles smaller than 1 μm have been successfully encapsulated, the product is larger than 100 μm. The excellent drying in this process also means that it is not the method of choice when encapsulating volatile materials, although this can sometimes be accomplished if the volatile is trapped in the coating or a carrier particle. The process cannot coat fluids without first converting them into solid particles.

The need for a license to use the process commercially is often regarded as a draw-

TABLE 7

Various Sizes and Capacities of
Wurster Equipment

Model (in.)	Estimated load (kg)
4/6	0.5—1.0
6	1.0—3.0
12	12—25
12/18	20—40
18	40—80
18/24	80—120
24	120—160
46	300—450

back by individuals or companies who consider the royalties due the Wisconsin Alumni Research Foundation objectionable. In fact, the process is frequently more economical than other methods, even with the royalty. The royalties are payable until 1983.

V. EQUIPMENT

Wurster equipment is available from manufacturers (see Appendix 2) in a variety of sizes and configurations. The smallest readily available has a perforated plate 4 in. in diameter and a volume capacity of 0.05 ft³. This unit will coat approximately 1 lb (0.5 kg) of material, depending on its density. The largest unit available has a volume of 21 ft³ and will accommodate 1000 lb (450 kg) of most materials. Table 7 lists the various sizes and capacities of the units readily available. Other sizes can be designed on a custom basis.

All Wurster equipment is available with a variety of options. The simplest equipment is suited for the coating of large particles and is controlled by an operator. More sophisticated equipment can be used to coat very small particles as well as larger particles and can be automated to whatever degree is desired.

A. Laboratory Equipment

The laboratory units are usually designed to be very versatile, but because it is difficult to anticipate all the necessary variables in an experimental situation, they are usually not highly automated. The units are often shipped with the blower mounted as shown, but it is easily removable for remote location. Figure 8 is a schematic of such a unit.

The unit in Figure 8 is too large to be considered highly portable, although it can be pallet mounted and moved with minimal difficulty. A unit designed to be very portable is available. Two people are required to unpack it and set it up, but one person can move it from room to room. To make the unit this portable, some versatility was sacrificed, and this unit is best used on particles larger than 30 mesh. Both the unit in Figure 8 and the more portable unit will accommodate interchangeable coating chambers of either the 4/6 in. size or the 6 in. size.

Lab units are usually supplied with positive displacement gear pumps, but can be ordered with any type of pump desired. Frequently, the drive unit is designed to be used with several different types of pumps so that a wider variety of materials may be used for the coating.

Air atomizing nozzles are used in all units of this size because the spray rates used are too low for conventional hydraulic nozzles to work well.

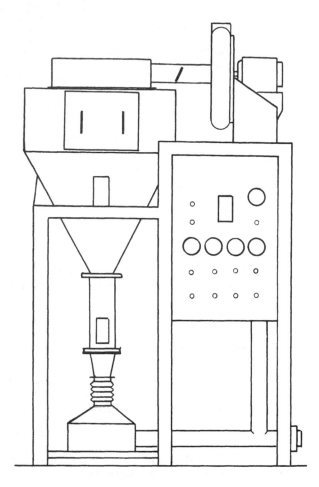

FIGURE 8.   Schematic of a Wurster unit for laboratory use.

Frequently, special features can be incorporated into a lab unit at the time of man-
ufacture if the purchaser is able to discuss the intended application. This can save time
in the lab by having a unit well suited to its intended task as well as considerable
expense in adding components to a unit after installation.

B. Manufacturing Equipment

Usually, a company contemplating the purchase of production equipment has had
the benefit of processing the material in question in a Wurster unit, either in their own
facility or in a lab offering such services (Appendix 3). This experience is extremely
valuable in selecting the necessary features and configuration for the production units.
Since the larger production equipment is frequently used for a single product, it is
often designed to be less versatile than the experimental equipment. If it is anticipated
that the unit will be used for a variety of products, it can be made as versatile as
desired, but this is necessarily more expensive.

The following specifications were prepared for a unit with two coating chambers,
one with a 12-in. plate diameter and the other with an 18-in. plate diameter. This unit's
capable of coating a wide range of particles.

Specifications for 12/18 in. Unit

General — 12 in. and/or 18 in. — All materials T-304 stainless steel with #4 finish; all inside and outside welds to be ground and polished to a #4 finish

Coating Chambers — 12/18 in. truncated and 18 in. straight chambers, each with three viewing windows located 120° apart

Nozzle — Air atomizing type, stainless steel construction, capable of atomizing 600 ml/min.

Partitions — One W12-6/9 for the 12/18 in. unit and one W18-8/12 for the 18-in. unit

Plate — W12-1 for the 12/18 in. unit and W18-6 for the 18-in. unit; each plate to be covered with a 325 mesh stainless steel screen

Bed design — Cart system with nozzle mounting housing

Plenum chamber — Air distribution type with one centrally located lift cylinder capable of 250 lb thrust at 80 psi with a 6-in. stroke; must operate at 100°C

Heat — Steam heat with output of 75,000 BTU/hr at 500 SCFM and 100°C temperature rise

Expansion chamber — Sized to fit filter system described below with air-actuated vibrators mounted to the chamber walls; three viewing windows in the inclined walls, 120° apart; two viewing windows on the access doors

Window material — Lexan® MR-4000

Frame — All stainless steel construction with the front legs recessed for easy access to process area

Filters — 16 quantity 30-in. bags rated at 667 SCFM; continuous-cleaning type with 99.9% efficiency above 0.5 μm

Pump and drive — A positive displacement metering pump with teflon seals rated at 100 to 1000 ml/min based on drive output; drive to be a digital metering unit, constant torque, reversible drive with a revolution per minute readout; motor speed controlled within 1/4% of set point, unit to be explosion proof

Blower — A pressure blower capable of 750 SCFM at 25-in. water pressure, to be spark proof with an explosion-proof motor

Control Panel — All stainless steel construction to be protected from explosion and vibration; this panel to include all the following controls to meet the following specifications:

1. Start button control station
2. Stop button control station
3. Air flow monitor alarm, 0 to 1000 CFM
4. Inlet temperature recorder/alarm, 0 to 150°C range
5. Temperature control controlling steam heat, reliable to ± 2.5°C with low air flow safety interlock, 0 to 200°C range; digital control with zero-center deviation meter
6. Outlet temperature recorder/alarm, 0 to 150°C range
7. Forward and reverse control stations for the pumping system
8. Five-digit elapsed-time indicator with manual reset
9. Three-position switch for pump and timer control, positions to be manual/off/automatic; pump on manual, pump and timer on automatic mode
10. Plenum raise/lower valve — double acting
11. Plenum raise/lower pressure gauge — 0 to 100 psi
12. Plenum raise/lower pressure regulator
16. Atomizing air pressure gauge/alarm — solid state indicating control, 0 to 100 psi
17. Atomizing air pressure regulator — precision type
18. Atomizing air flow readout/alarm
19. Process air flow valve pressure gauge — 0 to 30 psi

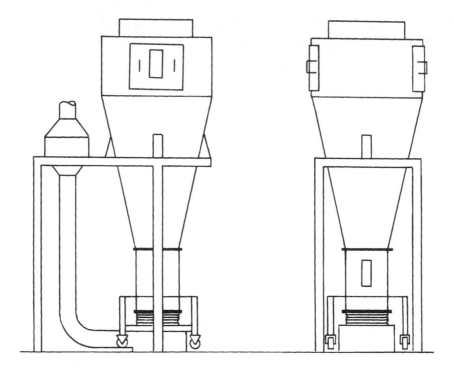

FIGURE 9.   View of an 18 in. unit.

20. Process air flow valve pressure regulator—precision type
21. Panel pressurization gauge/ alarm — solid state indicating control, 0 to 30 psi
22. Panel pressurization regulator — general purpose
23. Filter air supply on/off valve
24. Filter pressure gauge— 0 to 100 psi
25. Filter pressure regulator — general purpose
26. Filter pulse on/off
27. Filter timer and controls
28. Pump RPM
29. Liquid flow monitor/alarm
30. Liquid nozzle monitor/alarm — solid state indicating control, 30 to 0 to 30 psi.
31. Unit inlet compressed air pressure/alarm — solid state indicating control, 0 to 160 psi
32. Air vibrator on/off valve
33. Air vibrator pressure gauge — 0 to 100 psi
34. Air vibrator pressure regulator — general purpose

Figure 9 is a view of an 18-in. unit designed to coat small particles.   As mentioned at the beginning of this chapter, all of the functions are monitored and controlled from the control panel (Figure 10). The following is a brief description of the controls and their functions.

Number 1 and Number 2 are control stations which activate or deactivate an AC magnetic contactor which controls the blower motor. Number 5, the temperature controller, is also controlled, being on when the blower is running and off when the blower is not running. Number 3 is an airflow monitor with optional alarm. This monitor, in conjunction with an annubar or pitot tube, will give process air flow volumes through the unit. The readouts can be in inches of static pressure, a magnehelic or eagle eye,

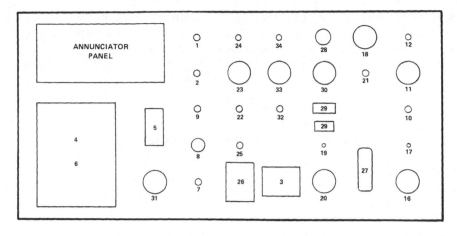

FIGURE 10.   Control panel for a Wurster unit.

or direct readout in SCFM. The airflow through the system is controlled by Number 18, gauge readout, and Number 19, pressure regulator. These control the compressed air used to change the setting on the damper, increasing or decreasing the air allowed through the system. Number 4 and Number 6 are inlet and outlet temperature readouts—optional recorder and/or alarms. The inlet temperature monitors the temperature set by the temperature controller, Number 5. The outlet temperature monitors the bed temperature or the temperature to which the material being coated is subjected. Number 7 is a control station which activates an AC magnetic contactor or directly turns on the pumping system. Many units are supplied with reversing contacts, making it possible to reverse the pump from the panel. Others have reversing levers on the pump drives. Number 8 is an elapsed time indicator. This is used to show the total elapsed time during a run. Number 9 is a selector switch allowing for manual or automatic control of the pump. If Number 9 is in the automatic mode, the pump and timer will run only when the blower is on. In the manual mode, the pump will run even if the blower is off. Numbers 10, 11, and 12 operate the plenum or cart raise/lower. This allows for easy removal of the product container for loading or unloading, but makes a tight seal to minimize leaks when the system is compressed for operation. Numbers 13, 14, and 15 are necessary only if the unit is supplied with a nozzle having an external control for shutting off the liquid flow. Numbers 16 and 17 control the pressure of the air used for the atomizing nozzle. Number 16 may have an optional alarm. This compressed air may also be monitored by option Number 27, airflow meter readout/ alarm. Numbers 20 and 21 vary, depending on the electrical code requirements set by the purchaser. At minimum, all units are supplied with pressurized panels. Numbers 22, 23, 24, 25, and 26 control the filtering system used to separate the product from the exhausting air. Number 26 has variable pulse duration and frequency timers which are easily adjusted. Number 28, an option or standard with some pumping systems, allows for visible readout of pump output. Some systems even allow adjustment of the pump output from the panel. Number 29, an option with or without alarm, reads the line pressure of the liquid feed supplied to the unit through a sanitary diaphragm. Number 31, an option with or without alarm, monitors the incoming pressure to the panel so as to minimize problems caused by outside air failure. Numbers 32, 33, and 34, options recommended for fine particle work, help to keep finer material from clinging to chamber walls. Number 35, an option, is a monitor and/or control for regulating the percent of humidity in the processing air.

The above is a description of the most common controls used, but it is not inclusive. Other items, such as solvent monitors, annunciator panels, and differential pressure monitors, can be installed to help monitor the coating operation.

Further information regarding the process and its many applications, as well as assistance in selecting equipment for specific applications, is available from the authors.

APPENDIX 1

Coatings successfully applied by Coating Place, Inc., Verona, Wis.

Acid soluble resins
Acrylics
Acrylonitrile polymers
AEA Sankyo
Alginates
Base-soluble resins
Candelilla wax
Capsul
Carageenen
Carbowax®
Carboxymethyl cellulose (CMC)
Casein
Castorwax®
Cellulose acetate (CA)
Cellulose acetate butyrate (CAB)
Cellulose acetate phthalate (CAP)
Chitosan
Chlorinated rubbers
Clays
Corn syrup solids
Crushed limestone (w/binder)
Dextrins
Enzymes
Epoxy resins
Ethyl cellulose
Ethyl methacrylates
Eudragit resins (types E, L, R, and S)
Gelatins
Glycerides
Glycols
Gum Arabic
Hydrogenated fats and oils
Hydroxyethyl cellulose
Hydroxypropyl cellulose (HPC)
Hydroxypropyl methyl cellulose (HPMC)
Hydroxypropyl methyl cellulose phthalate (HPMCP)
Hydrocarbon resins
Kel-F
Lard
Methyl methacrylates
Methyl cellulose (MC)

Microcrystalline waxes
Milk solids
Molasses
Nitrocellulose
Nylon
Paraffin waxes
Phenolic resins
Polyethylene
Polyvinyl acetate (PVA)
Polyvinyl alcohol (PVAL)
Polyvinyl chloride (PVC)
Polystyrene
Polyvinylacetate phthalate (PVAP)
Polyvinylidene chloride (PVDC)
Polyvinyl pyrrolidone (PVP)
Proteins
Rosin esters
Rubbers
Shellac
Starches (modified)
Starches, amylose
Stearic acid
Sucrose
Sugars (other than sucrose)
Surfactants
Synthetic waxes
Teflon®
Vermiculite® (powdered w/binder)
Whey
Zein

APPENDIX 2

Manufacturers of Wurster equipment*

Lakso Company
Box 929
Leominster, Massachusetts 01453

APPENDIX 3

Research and development laboratories using the Wurster process

Coating Place, Inc.
P.O. Box 248
Verona, Wisconsin
608-845-9521

* Licensees of the Wurster process are free to obtain equipment from anyone they select.

REFERENCES

1. Wurster, D. E., Means for Applying Coatings to Tablets or Like, U.S. Patent 2,799,241, 1957.
2. Wurster, D. E., Granulating and Coating Process for Uniform Granules, U.S. Patent 3,089,824, 1963.
3. Lindloff, J. A., Apparatus for Coating Particles in Fluidized Bed, U.S. Patent 3,117,027, 1964.
4. Wurster, D. E., Apparatus for Encapsulation of Discrete Particles, U.S. Patent 3,196,827, 1965.
5. Wurster, D. E., Process for Preparing Agglomerates, U.S. Patent 3,207,824, 1965.
6. Wurster, D. E., Particle Coating Apparatus, U.S. Patent 3,241,520, 1966.
7. Wurster, D. E., Particle Coating Apparatus, U.S. Patent 3,253,944, 1966.
8. Hall, H. and Hinkes, T., Air suspension encapsulation of moisture sensitive particles using aqueous systems, in *Microencapsulation: Processes and Applications,* Vandergaer, J. E., Ed., Plenum Press, New York, 1974, 145.
9. Sarnotsky, A. A., Evaporation of solvents from paint films, *J. Paint Technol.,* V 41, 692, 1969.
10. Crowley, J. D., Teague, G. S., Jr., and Lowe, J. W., Jr., A three-dimensional approach to solubility, *J. Paint Technol.,* 38, 269, 1966.
11. Anon., Computer directs tablet making at new plant, *Chem. Eng. News,* October 31, 5, 1977.
12. Pondell, R. E., Enteric Coating Study using CAP, In-house Report, Coating Place, Inc., Verona, Wis., July, 1973.
13. Wurster, D. E., Singiser, R. E., and Lowenthal, W., Enteric filmcoats by the air suspension coating technique, *J. Pharm. Sci.,* 50(2), 168, 1961.
14. Wurster, D. E., Coletta, V., and Rubin, H., Wurster coated aspirin. I, *J. Pharm. Sci.,* 53 (8), 953, 1964.
15. Wood, J. H. and Syarto, J., Wurster coated aspirin. II, *J. Pharm. Sci.,* 53(8), 877, 1974.
16. Theeuwes, F., Elementary osmotic pump, *J. Pharm. Sci.,* 64, 1987, 1975.
17. Debbie, J. C., Recent Advances in Rabies Vaccination, presented at Symp. on Recent Advances in Rabies, Atlanta, September 1976 and American Public Health Assocation, Miami, October 1976.
18. Sparks, R. E., Mason, N. S., Goldenhersh, K., and Huang, W., Microcapsules for augmenting artificial kidney function, in *Microencapsulation,* Vol. 3, Nixon, J. R., Ed., Marcel Dekker, New York, 1976, chap. 9.
19. Sparks, R. E., Mason, N. S., Goldenhersh, K., and Huang, W., Rationale for the Use of Microencapsulated Sorbents in Uremia and Other Illnesses, presented at the Conf. on Plastics in Medicine and Surgery, Glasgow, Scotland, September 1975.
20. Hall, H. and Hinkes, T., Wurster process for controlling pesticides, in *Proc. 1976 Int. Controlled Release Pesticide Symp.,* Cardarelli, N. F., Ed., University of Akron, Ohio, 1976, 4.1.
21. Abrams, J. and Hinkes, T., Encapsulation of solid particles, in *Proc. 1974 Int. Controlled Release Pesticide Symp.,* Cardarelli, N. F., Ed., University of Akron, Ohio, 1974, 7.1.
22. Hall, H., Controlled application of fungicide onto onion seed, in *Proc. 1975 Int. Controlled Release Pesticide Symp.,* Harris, F. W., Ed., Wright State University, Dayton, Ohio, 1975, 393.
23. Hall, H. and Hinkes, T., Case histories of controlled release products, *Proc. Am. Chem. Soc. National Meeting,* Rep. 14, Chemical Marketing and Economics Division, San Francisco, Calif., September 1976, 202.
24. Abrams, J., Acceptability and performance of encapsulated warfarin, *Pest Control,* 14, May, 1974.
25. Hall, H. and Pondell, R., Controlled Release of Pesticides, presented at Chemical Specialties Manufacturers Association meeting, Chicago, May 1977.
26. Investigations in Fish Control, No. 25—28, U.S. Department of Interior, Fish and Wildlife Service, Washington, D. C., 1969.
27. Hinkes, T., Seed Coating Composition and Coated Seed, U.S. Patent 3,950,891, 1976.
28. Fuller, S. R., Jr. and Munzel, H. E., Electrophotographic Carriers, U.S. Patent 3,725,118, 1971.
29. Queener, C. A., Ralston, W. G., Smith, T. C., and Welsh, J. P., Improved Electrophotographic Process, U.S. Patent 3,778,262, 1971.
30. Kulka, W. J. and Munzel, H. E., Electrophotographic Carriers, U.S. Patent 3,798,167, 1971.
31. Coating and Specialty Products Department, Report from the Coatings Service Laboratory, CSL-204A, Hercules, Inc., Wilmington, Del., 19899.

Chapter 8

MICROENCAPSULATION USING PHYSICAL METHODS

George R. Somerville and John T. Goodwin

TABLE OF CONTENTS

I. GENERAL

It has been well established that encapsulation can be an effective means of conveniently handling toxic or noxious materials and, in many cases, of providing a controlled release mechanism. Rather than to review specific controlled release formulations, it is the purpose of this chapter to describe several processes that might be applied to the encapsulation of pesticides and to discuss their versatility and limitations with respect to possible capsule shell compositions.

The encapsulation techniques under consideration fall under the general heading of physical processes in that nozzle devices are involved, although to categorize them as such might be misleading, as chemical processing is frequently utilized. One characteristic that these several physical processes have in common is that fluid shell formulations are used, and the shell is hardened after the capsules are formed. Basic criteria for selection of the fluid shell formulation are that it must be nonreactive and essentially immiscible with the material being encapsulated and that it must be capable of being rapidly hardened to form a film. Shell formulations suitable for these processes include solutions, latexes, and hot melts.

The criterion of nonreactivity of the shell formulation with the material being encapsulated is quite obvious, although a slow, limited reaction may be permissible in some instances. For example, a crosslinking of the polymeric shell by core components may be tolerated or, in fact, be desirable. The immiscibility requirement can, in general, be met without difficulty, provided the hardened shell material meets the specifications for the end use of the capsule, e.g., resistance to its storage, handling and use environments, and the release of its contents under prescribed conditions. For example, if a water-insoluble material is to be encapsulated, an aqueous shell solution or a latex would be appropriate. Conversely, water or an aqueous solution would be encapsulated in a nonaqueous shell formulation. In some cases in which both the core and shell materials are handled as hot melts during the encapsulation process, it is possible to circumvent the immiscibility requirement because the shell is hardened before a significant degree of intermixing occurs.

The shell hardening mechanism is dependent upon the type of shell formulation employed. The general mechanisms are chemical reaction, solvent extraction or evaporation, cooling, and combinations of these general mechanisms. For example, an aqueous sodium alginate or sodium polypectate solution is rapidly converted to the respective calcium salt by receiving the freshly formed capsule in an aqueous calcium salt solution, after which the capsule is rinsed and the residual water removed. Similarly, latexes or aqueous polymer solutions may be combined with sodium alginate or polypectate and the capsules received in a calcium salt solution. In this case, the insoluble calcium alginate or polypectate serves as a gelling agent to give the capsule initial physical integrity. Following a water rinse, the capsule is dried, and the primary component forms a continuous film. Fluidized bed drying is a particularly useful technique for residual water removal, although overdrying of some shell systems can lead to excessive shell brittleness.

Shell polymers dissolved in volatile solvents may be hardened by evaporation of the solvent as the capsules are projected from the encapsulating nozzle. Depending upon the solvent volatility, this approach may be practical for only relatively small capsules, as reasonable trajectory limitations preclude the drying of large capsules in using this technique. Solvent extraction, on the other hand, can be achieved by receiving the freshly formed capsules in an appropriate nonsolvent. With this technique, the solvent-nonsolvent system must be such that polymer precipitation does not occur, and shells which go through a tacky state during solvent extraction must be kept physically separated.

Regardless of the hardening mechanism employed, shell solutions and latexes offer the advantage of providing an excellent means of achieving high capsule payloads because of the shell shrinkage which takes place as the solvent is removed. The shell formulation viscosity required for effective operation of the nozzle devices may be achieved by adjustment of the total solids or, in some cases, by selection of a polymer in the proper molecular weight range. If too great a total solids content is required to attain the required viscosity, a compatible gelling agent might be incorporated in the formulation with a reduction in total solids.

Cooling as a shell hardening mechanism, if applicable, is the simplest system from an operational standpoint. In those cases where the capsule is formed from a molten shell system, the shell is merely cooled to the solidification point as the capsule is projected from the encapsulating nozzle, and provision need not be made for chemical treatment, drying, or solvent recovery. It should be noted, however, that higher capsule payloads can be achieved using shell solutions and latexes than with molten shell systems, as in the former cases, a significant shrinkage of the shell occurs upon drying. In selecting a molten shell formulation, attention should be directed toward avoiding formulations which have supercooling characteristics, as supercooling can present problems in achieving rapid hardening after the capsule is formed. This might be overcome by the use of additives in the molten shell.

A variation of cooling as a hardening technique involves the use of a heated polymer solution containing an agent which forms a gel as the temperature is lowered; thus, the required degree of structural integrity is provided while the residual shell solvent is removed.

The various encapsulation processes to be described require that the material to be encapsulated be pumped through the nozzle devices. Thus, liquids are ideally suited to these processes. In some cases, it is appropriate to encapsulate finely divided solids which are slurried in a liquid vehicle; this technique has been useful in controlled release applications by taking advantage of the physical properties of the core matrix, particularly in the case of a solid which serves as the liquid vehicle while in the molten state, to provide the desired release pattern. It has been demonstrated that the core matrix composition of such capsules can have a substantial influence on release rate.[1] The liquid slurry vehicle might also serve to isolate the solid particles from the fluid shell in those cases where direct contact should be avoided until the shell is hardened. Finally, some solid core materials can be encapsulated in the molten state, as liquids. Obviously, the constraint of thermal stability is imposed.

The choice of shell materials meeting the general requirements of the physical encapsulation processes is quite broad and includes many commercially available synthetic polymers as well as natural gums, waxes, and resins. More often than not, mixtures of shell materials are employed for the purpose of achieving encapsulation, e.g., the use of a gelling agent to provide initial structural integrity, or of achieving the desired shell properties. Postencapsulation treatment may also be appropriate, particularly with respect to controlled release applications. For example, the degree of swelling of moisture-sensitive capsule shells can often be regulated by crosslinking of the shell polymer. One point that should be borne in mind in the development of a controlled release shell formulation is that of stability with respect to release characteristics. Release rates may change with time as the result of such phenomena as continued crosslinking or the occurrence of polymorphs in the shell or core matrix.

Plasticized shells might also be desirable in the end product where thin, brittle shells lead to excessive capsule rupture during handling and use. Plasticization might be ac-

INTERNAL PHASE

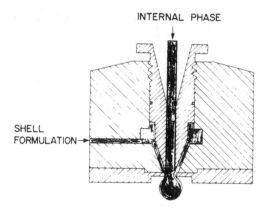

SHELL
FORMULATION→

FIGURE 1.   Gravity-flow nozzle. (From Goodwin, J. T. and Somerville, G. R., *Microencapsulation Processes and Applications*, Vandegaer, J. E., Ed., Plenum Press, New York, 1974, 155. With permission.)

complished by either incorporation of the plasticizer in the fluid shell formulation or postencapsulation treatment.

II. MICROENCAPSULATION PROCESSES

Turning now to the processes themselves, initial work at Southwest Research Institute on the development of nozzle-type encapsulation processes and equipment resulted in a gravity-flow device represented cross-sectionally in Figure 1.[2] The fluid shell material enters the port on the side of the nozzle device and flows through an annular space to a circular orifice located at the base of the device where it forms a membrane. The material to be encapsulated drops through a center tube and onto the membrane where it builds up and causes the membrane to distend. Eventually, a compound fluid droplet breaks loose and falls from the nozzle. Surface tension causes the fluid capsule to assume a spherical shape as it falls to an appropriate hardening bath. As the capsule breaks loose from the nozzle, the membrane reforms across the orifice, and the next capsule begins to take shape.

Following this initial development, interest arose in seamless capsules which were much too small for gravity-flow formation, i.e., about 400 to 500 microns in diameter, and centrifugal force was utilized to assist in forming the capsules. The multiorifice centrifugal head illustrated in Figure 2[3] is representative of several similar devices which have been developed. The principal element is a head which rotates about a vertical axis. The fluid shell material enters the head in two streams, each being directed into an internal groove where centrifugal force holds the material until it overflows an internal weir and enters an area containing a row of counterbored orifices. Membranes form across each of the orifices in a manner similar to the performance in the previous device. The material to be encapsulated is fed onto a concentrically rotating disc which atomizes the core material and projects it onto the membranes. Capsule formation takes place when the combined mass of core and shell material becomes sufficient for centrifugal force to overcome the cohesive force of the shell material, at which point a capsule is projected outwardly from the periphery of the head.

For any given head configuration, then, the capsule size is indirectly proportional

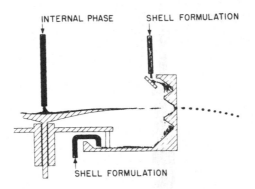

INTERNAL PHASE   SHELL FORMULATION

SHELL FORMULATION

FIGURE 2. Multiorifice centrifugal head. (From Goodwin, J. T. and Somerville, G. R., *Microencapsulation Processes and Applications,* Vandegaer, J. E., Ed., Plenum Press, New York, 1974, 155. With permission.)

FIGURE 3. Sloped-face multiorifice head. (From Goodwin, J. T. and Somerville, G. R., *Microencapsulation Processes and Applications,* Vandegaer, J. E., Ed., Plenum Press, New York, 1974, 155. With permission.)

to the rotational speed of the head. However, size control through speed adjustment has its limitations; the rotational speed can be reduced just so far in making larger capsules before gravitational force becomes significant, resulting in either poor capsule formation or a dribbling of the capsule materials down the outer surface of the head. This problem was solved by sloping the outer face with the declination of the orifice axes being the resultant of the horizontal centrifugal force vector and the vertical gravitational force vector. The principal element of this device is shown in Figure 3,[4] a 90-orifice head of 8-in. diameter.

As the foregoing devices provided a capability of controlling capsule size and uniformity, development effort shifted to the problem of increasing the rate of production. This effort emphasized the combination of two techniques: the earlier centrifugal encapsulation and fluid extrusion. Figure 4 represents encapsulation by means of simple fluid extrusion through a fixed concentric-tube nozzle. The core material is pumped through the central tube and the shell material through the annular space to extrude a

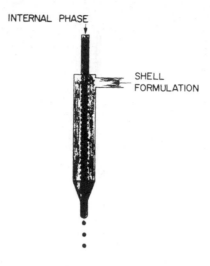

FIGURE 4. Simple extrusion. (From
Goodwin, J. T. and Somerville, G. R.,
*Microencapsulation Processes and Appli-
cations*, Vandegaer, J. E., Ed., Plenum
Press, New York, 1974, 155. With per-
mission.)

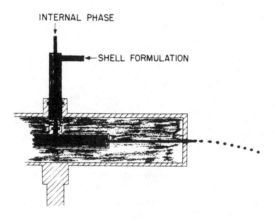

FIGURE 5.   Multiorifice centrifugal extrusion head (From
Goodwin, J. T. and Somerville, G. R., *Microencapsulation
Processes and Applications,* Vandegaer, J. E., Ed., Plenum
Press, New York, 1974, 155. With permission.)

fluid "rod" of the core encased in a sheath of shell material. At some point beyond
the tip of the nozzle, the compound fluid "rod" breaks into individual fluid capsules.
Such a device enjoys a much higher rate of capsule formation than do the previously
described centrifugal devices where capsule-by-capsule formation occurs; however, the
uniformity of the capsules is relatively poor, and size control is somewhat limited. A
combination of the features of the two types of devices resulted in the centrifugal
apparatus shown schematically in Figure 5. The encapsulation head rotates about a
vertical axis with a concentric feed tube entering through a seal arrangement. The core
material is pumped through the central feed tube into an inner chamber and then flows
through radially disposed tubes which penetrate orifices about the periphery of the

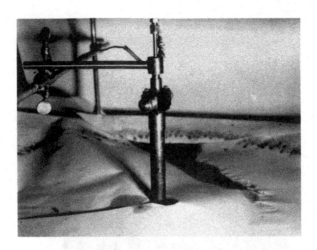

FIGURE 6. Centrifugal extrusion device.

rotating head. The shell formulation is pumped through the outer feed tube and flows through the annuli created by the sets of radial tubes and orifices which comprise the individual nozzles. As the head rotates, nodes form on the extruded "rods" and eventually break off as individual capsules. In actual practice, threaded concentric-orifice spray-type nozzles are used in the head which is appropriately ported to direct the shell and core streams to the respective orifices. Easy removal and replacement of the nozzles facilitates cleaning, maintenance, and conversion to produce a different capsule. In selecting the concentric-orifice nozzles for producing a particular capsule, the ratio of shell formulation to core material is an important consideration, as it determines the relative pumping rates of these two streams. Ideally, the cross-sectional areas of the two flow paths at the nozzle tip will be such that both streams are extruded at substantially the same velocity, and a smoother interface is provided.

A dramatic increase in production rate was realized by this process. Capsules of 350 μm diameter have been produced at rates exceeding 300,000/sec per orifice. This is in comparison to rates of 20 to 30/sec per orifice for the previously described centrifugal devices. Additionally, the totally enclosed system simplifies temperature control and operation under controlled atmospheres and permits encapsulation of volatile liquids without the losses which are incurred when using an open feed disc. Capsule size increases with feed rate and decreases with rotational speed. Thus, capsules of the desired size can be produced by balancing these two factors. Figure 6 is a photograph of a single-nozzle laboratory device. Note the formation of nodes on the extruded "rod" and the eventual break-up into individual capsules.

For those systems involving delicate fluid shells which have a tendency to break or seriously deform on impact with a hardening bath, a submerged extrusion nozzle was developed.[5] Shown in Figure 7, this device consists of a concentric-tube extrusion nozzle which is mounted in a duct which tapers to a reduced diameter just beyond the nozzle tip. As with a simple extrusion nozzle, the core material is pumped through the central tube and the shell formulation through the annulus. An inert, immiscible carrier fluid flows through the duct, and as the velocity increases with reduced duct diameter, the extruded "rod" is attenuated and finally breaks into individual capsules which are subsequently hardened. For a given nozzle configuration, capsule size is determined by the difference in velocities of the extruded "rod" and the carrier fluid, i.e., the relative pumping rates. To cite an example of operation, this device may be

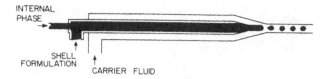

FIGURE 7.   Submerged extrusion nozzle. (From Goodwin, J. T. and Somerville, G. R., *Microencapsulation Processes and Applications*, Vandegaer, J. E., Ed., Plenum Press, New York, 1974, 155. With permission.)

FIGURE 8.   Vibrating extrusion nozzle.

used for the encapsulation of water in a molten, waxy shell material. In this case, the carrier fluid is hot water, with the temperature in the duct maintained just above the shell solidification point. Capsule hardening is accomplished by jacketing the discharge tube with cooling water to lower the carrier fluid temperature to just below the shell solidification point. The hardened capsules are removed from the carrier fluid stream by a screen or other mechanical means, and the carrier fluid passes through a heater and back to the inlet to form a closed loop.

It should be noted that the submerged nozzle device can be operated in a vertical position with a relatively low flow rate of carrier fluid to provide buoyancy that allows formation of capsules larger than can be made with the original gravity-flow device.

The size of capsules produced by the foregoing physical processes typically falls in a more or less normal distribution about the mean, and for most applications, the overall size range is acceptable. In some instances, however, a very narrow size range may be required and cannot be attained without the uneconomical practice of sieving out oversize and undersize fractions. If an extremely narrow size distribution is mandatory, the extrusion nozzle devices — both rotating and fixed — can be modified to incorporate a vibration source to impose uniformity of node formation on the extruded "rods" and thereby result in capsules of excellent size uniformity. Figure 8 is a photograph of a vibrating nozzle. Randomly selected samples have exhibited a maximum

FIGURE 9.   Vibrating centrifugal extrusion nozzle.

weight variation from the mean of about 2 wt %. Figure 9 is an experimental dual centrifugal nozzle equipped with a piezoelectric stack as the vibrating source.

III. PRODUCTION FACILITIES

The foregoing nozzle-type devices have been used in the laboratory for a number of years, primarily for experimental product development. Typically, the laboratory devices contain from one to four nozzles. For the production of test quantities of capsules using the laboratory devices, extended runs are required with an attendant high labor cost per pound of capsules. The scale-up potential of these devices has been demonstrated, however, and maintenance problems have been minimal, being restricted primarily to such items as seals and bearings. With the centrifugal devices, scale-up is effectively achieved by increasing the number of nozzles through a greater head diameter and/or multiple rows of nozzles. The capacity of the submerged nozzle device is increased by manifolding a multiplicity of nozzles.

Production facilities which have been constructed and operated include a centrifugal extrusion system designed for handling a molten shell and having a demonstrated capacity in excess of 90,000 lb/month on single-shift operation and one designed for a chemically hardened shell system with a capacity of 10,000 lb/month. A manifolded submerged nozzle facility has been successfully operated at a level of about 25,000 lb/month.

Figure 10 is a view of the hot-melt centrifugal extrusion device mounted on a platform above a plastic-lined collection cone. The capsules, which are hardening by air-cooling in their trajectory from the head, flow from the cone to a vibrating conveyor that transports them to a sieve for classification prior to packaging.

As to the question of capsule manufacturing costs, the capital equipment investment is generally quite modest, the magnitude depending upon such factors as the type of equipment utilized, postencapsulation treatment requirements, production level, and capsule composition. Material costs vary over a wide range, depending primarily upon

FIGURE 10.   Hot-melt encapsulation system.
(From Goodwin, J. T. and Somerville, G. R.,
*Microencapsulation Processes and Applica-
tions*, Vandegaer, J. E., Ed., Plenum Press,
New York, 1974, 155. With permission.)

capsule payload and shell composition. In general, labor is the significant factor in
the capsule manufacturing cost at low production levels and decreases at higher production levels with materials costs becoming dominant, particularly with low capsule
payloads and relatively expensive shell ingredients.

Although the ability to encapsulate a given material using the physical processes
does not necessarily mean that the capsule performance specifications can be met, the
broad spectrum of available shell material combinations enhances the possibility of
obtaining an acceptable capsule. Technical feasibility must be evaluated on a case-by-
case basis, and it may be necessary to make trade-offs in the specifications in arriving
at the optimum capsule.

REFERENCES

1. Somerville, G. R., Goodwin, J. T., and Johnson, D. E., Controlled release of quinidine sulfate
   microcapsules in *Controlled Release Polymeric Formulations*, Paul, D. R. and Harris, F. W., Eds.,
   ACS Symp. Series 33, American Chemical Society, Washington, D.C., 1976, 182.
2. Raley, C. F., Burkett, W. J., and Swearingen, J. S., U.S. Patent 2,766,478, 1956.
3. Somerville, G. R., U.S. Patent 3,015,128, 1962.
4. Somerville, G. R., U.S. Patent 3,310,612, 1967.
5. Somerville, G. R., U.S. Patent 3,389,194, 1968.

Chapter 9

CONTROLLED VAPOR RELEASE FROM HOLLOW FIBERS: THEORY AND APPLICATIONS WITH INSECT PHEROMONES

Thomas W. Brooks

TABLE OF CONTENTS

I. INTRODUCTION

Controlled release of vaporizable materials from hollow fibers has been under investigation and commercial development since 1974 and for certain applications has shown considerable potential.[1-3] The hollow fiber concept of controlled vapor dispensing shares technical and commercial objectives with other controlled release vehicles and systems, while possessing advantages and limitations which are more or less unique. This chapter will deal with the mass transport theory involved in hollow fiber controlled release vehicles, general features of hollow fiber controlled release devices, and practical applications of the concept to pest control with insect pheromones. While the basic concept of controlled vapor release with hollow fibers is potentially applicable to a variety of vaporizable active materials, the practical work discussed in this chapter will focus exclusively on insect pheromones, since this is the area where hollow fiber vapor dispensing has received greatest attention.

II. MASS TRANSPORT THEORY FOR VAPOR RELEASE FROM HOLLOW FIBERS

If a material is allowed to evaporate from the lumen of a hollow fiber sealed at one end and open at the other, the release curve obtained by plotting mass released vs. time is characterized by an initial steep slope followed by an extended lower slope flat region which approximates zero order release kinetics. This release behavior follows suit with all vaporizable materials as long as they are single component or comprised of mixtures of components with comparable volatilities. A model release curve is displayed in Figure 1. At any given temperature, release rates can be manipulated by adjustments in fiber internal diameter, or if a cluster of fibers is taken as a release point, the number of fiber ends in the cluster. The active life or longevity of a hollow fiber vapor dispenser is a function of fiber length, i.e., the length of the column of active material in the lumen reservoir at time zero.

A. Mechanism of Release

The mechanism for dispensing volatile materials from capillary channels is comprised of three steps as depicted in Figure 2:

1.   Evaporation at the liquid-vapor interface
2.   Diffusion from the liquid-vapor interface to the open end of the lumen
3.   Convection away from the open end

Generally, the diffusion step will be the rate controlling factor. The diffusive process can be likened to membrane release systems if the stagnant vapor column is viewed as a gaseous membrane through which active material permeates.[4] An important distinguishing feature of the gaseous membrane as contrasted with solid membrane media, however, is the absence of swelling, sorption-desorption effects, surface effects, and other phenomena which can influence release behavior of solid membrane-regulated release systems.

B. Mathematical Analysis and Transport Equations

The rate of diffusion of vapor in a capillary channel can be predicted by use of transport equations. The molar flux (Figure 3), N_a, is given by (Reference 5, p.522)

$$N_a = - \frac{cD}{1 - x_a} \frac{dx_a}{dx} \tag{1}$$

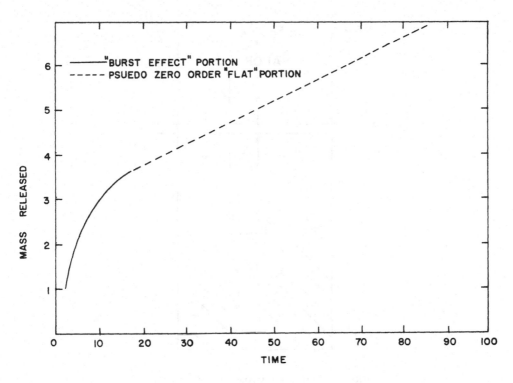

FIGURE 1.   Release curve for evaporation from the lumen of a hollow fiber sealed at one end.

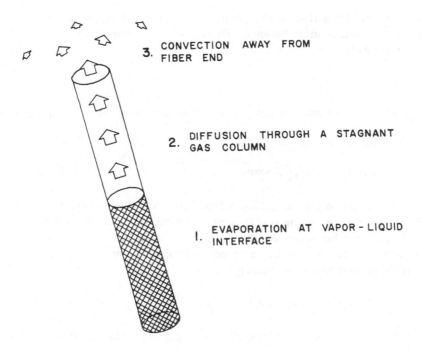

FIGURE 2.   Mechanism of vapor release from hollow fibers (one end sealed).

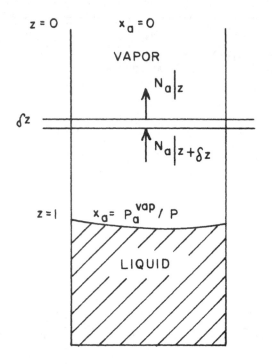

FIGURE 3.   Diffussion through a stagnant gas layer.

where x_a is the mole fraction of the volatile material, c its molar density, and D the diffusion coefficient. A mass balance, $dN_a/dz = 0$, combined with Equation 1 and assuming ideal gas behavior gives:

$$-\ln (1 - x_a) = C_1 z + C_2 \qquad (2)$$

where C_a and C_a are constants which must be determined by the boundary conditions.

$$\text{B.C. 1 at } z = 0, x_a = 0 \qquad (3)$$

$$\text{B.C. 2 at } z = 1, x_a = P_a^{vap}/P \qquad (4)$$

The first boundary condition indicates that there is zero concentration at the open end of the fiber, i.e., convection is sufficient to remove the volatile material away from the fiber. The second boundary condition implies vapor-liquid equilibrium at the liquid-vapor interface, and P_a^{vap} is the vapor pressure of the volatile material.

Applying these boundary conditions gives

$$(1 - x_a) = (1 - P_a^{vap}/P)^{z/\ell} \qquad (5)$$

The liquid-vapor interface is continuously moving away from the fiber open end, and the rate of this movement can be related to the molar flux, N_a.

$$-N_a = \frac{p}{M} \frac{d\ell}{dt} \qquad (6)$$

where P is the liquid density, M the molecular weight, and l is the distance between the fiber open end and the liquid-vapor interface.

Combining Equations 1, 5, and 6 gives

$$\frac{d\ell}{dt} = -\frac{cMD}{p\ell} \ln \frac{(1 - P_a^{vap})}{P} \qquad (7)$$

This then leads to

$$\frac{d\ell}{dt} = \left[\frac{-McD}{2p} \ln \left(1 - \frac{P_a^{vap}}{P}\right)\right]^{\frac{1}{2}} t^{-\frac{1}{2}} \qquad (8)$$

For most materials and temperatures considered $P_a^{vap} \ll P$. A Maclaurin expansion of the ln term gives

$$\ln \left(1 - \frac{P_a^{vap}}{P}\right) \approx -P_a^{vap}/P$$

so that

$$\frac{d\ell}{dt} = \left[\frac{McD}{2p} \frac{P_a^{vap}}{P}\right]^{\frac{1}{2}} t^{-\frac{1}{2}} \qquad (9)$$

Thus, the change in meniscus level with time, dl/dt, which is related to the dispensing rate, is predicted to be inversely proportional to the half power of time. Note that this relationship makes use of a pseudo steady state, i.e., a steady state is assumed at each instant even though the meniscus is continuously moving. The validity of this relationship was checked by measuring the movement of the meniscus for carbon tetrachloride. The results are shown in Figure 4 where the solid line represents Equation 9 and points are measured values for release rate. The agreement is quite good considering that the diffusion coefficient, D, was calculated by using an empirical equation. (Reference 5, Equation 16.3-1) The slope of the experimental curve is $-\frac{1}{2}$ as predicted by Equation 9.

The vapor pressure used in Equation 9 is not the true equilibrium vapor pressure of the material, but a reduced value due to the curved meniscus in the capillary. This reduction is a function of the radius of curvature of the meniscus[6] and as a result is dependent on the diameter of the hollow fiber. The curvature is also a function of surface tension and is therefore dependent on the nature of the material from which the hollow fiber is made. (In practice, a hollow fiber dispenser is best designed by making laboratory measurements using a hollow fiber of the size and material which will be used in the final product). The rate of release is dependent on temperature, since the material properties, P_a^{vap} D, and C are temperature dependent. The major contribution to this dependence is the vapor pressure and is given by the Clausius-Clapeyron equation.

$$\frac{d\ln P^{vap}}{dT} = \frac{\Delta H^{vap}}{RT^2} \qquad (10)$$

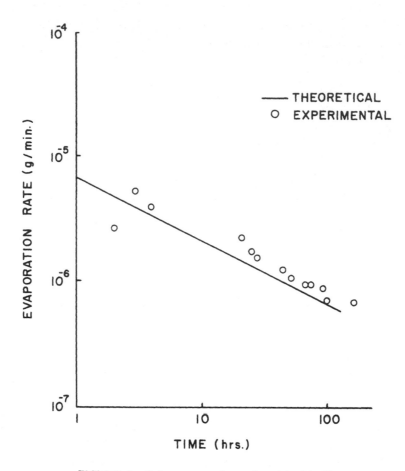

FIGURE 4.   Release curve for carbon tetrachloride.

A 30°F temperature change (e.g., from 70 to 100°F) will result in a vapor pressure increase for pheromone-type materials of about 200%, while the diffusivity, D, increases by only about 10%, and the molar gas density, c, decreases by about 5%. Thus, from Equation 9 it is noted that the dispensing rate will approximately double for a 30°F temperature rise.

C. Factors in Designing Hollow Fiber Controlled Release Formulations

Developing practical controlled vapor release formulations based on hollow fibers begins with specification of release rate and longevity for the individual dispenser. Basic design information is afforded by the release curve for the active material under consideration. Thus, for a given vaporizable material, hollow fiber controlled release dispensers can be formulated knowing only the release curve obtained with a particular fiber material and size at some specified temperature. Designing the dispenser to a specific release rate and lifetime is a simple matter of calculating the number of ends needed to give a desired dose rate and the length of fiber needed to achieve a prescribed life span.

The need to know the fiber material and internal diameter used in obtaining release curves deserves some explanation. It turns out that evaporation rates for many materials are sensitive to the diameter of the fiber lumen or capillary reservoir; that is, vapor flux is not directly proportional to the square of the capillary radius for all liquid and fiber material combinations as one might naively expect. Data in Table 1 serve to

TABLE 1

Comparison of Fluxes for Selected Materials from Two Different Hollow Fiber Sizes[a]

Material	Flux, cm/day[b]		Flux ratio 200 μm: 500 μ	Absolute release rate (μg/day/end)	
	200 μm I.D.	500 μm I.D.		200 μm	500 μm
Neutroleum-alpha[c]	1.8×10^{-2}	1.8×10^{-2}	1	5.9	37.4
Cubeb oil[d]	5.8×10^{-3}	1.5×10^{-2}	2.6	1.9	30.5
Multistriatin	2.1×10^{-2}	3.4×10^{-2}	1.6	6.8	69.6
4-methyl-3-heptanol	2.2×15^{-2}	3.16×10^{-2}	1.4	7.2	64.1

[a]   Fiber material is undrawn polyester (polyethylene terephthalate). All release measurements were made at $21 \pm 1°C$ and $65 \pm 5\%$ relative humidity.
[b]   The 200 μm and 500 μm fiber I.D.s correspond to lumen cross-sectional areas of 3.24 $\times 10^{-4}$ and 2.03×10^{-3} cm², respectively. This corresponds to a ratio of the larger to the smaller cross-sectional area of 6.3.
[c]   A proprietary general purpose deodorizer formulation comprised of a blend of essential oils and aromatic chemicals. Marketed by Fritzsche Bros., Inc., New York.
[d]   α-cubebene is a constituent which is extracted from the berry of the Java pepper plant.

illustrate this point. Steady state release rates measured on four different materials using 200 and 500 μm I.D. poly(ethylene terephthalate) hollow fibers are shown. Neutroleum-alpha is a commercial general purpose deodorizer formulation, while 4-methyl-3-heptanol (1), multistriatin (2), and α- cubebene (3) are compounds comprising the three component aggregation pheromone of an insect called the smaller European elm bark beetle, Scolytus multistriatus (Marsham).[7]

4—methyl—3—heptanol   Multistriatin   α-cubenene

Comparison of the fluxes of these materials from the two different sized fiber lumina reveals that release rate varies unpredictably with capillary diameter. Only with neutroleum-alpha is flux equal for the two diameters. With the three other materials, flux ratios ranged from 1.4 to 2.6. Thus, in going from the smaller to the larger diameter, release rate increased by more than the ratio of cross-sectional areas.

An explanation for observations such as these undoubtedly lies in the surface energy relationships between liquid charge and the lumen wall. It is known from classical physical chemistry that the vapor pressure above a liquid surface is dependent upon the radius of curvature of the surface.[6] For different liquid-fiber material combinations, the liquid meniscus-lumen wall contact angles will vary, as will radii of curvature of the menisci. It is important to recognize this phenomenon when considering the design and materials selection for hollow fiber controlled release formulations. In practice, such subtle complications can be avoided by measuring release curves with the fiber material and fiber internal diameter which will be used for a specific formulation. This is especially important with insect pheromones where biological dose-response relationships and economic considerations generally dictate rather precise release rates in order to make pheromone dispensers function reliably under field conditions.

For design purposes only, the flat portion of the release curve is used (see Figure 1). The initial high rate or "burst effect" portion of the curve can usually be ignored, since it represents only a small percentage of the total lumen charge and a relatively short period of total dispensing life. The burst effect portion of the release curve can oftentimes be eliminated in the formulation design by diluting the lumen charge with a high volatility inert material. This ploy is frequently useful in dispenser designs involving vapor mixtures where individual components differ substantially in volatility. Dilution can also be employed in making up short-lived dispensers which otherwise would have to be made in inconveniently short lengths.

Criteria for selection of fiber material will vary from one application to another, but generally the following considerations are involved:

1. Compatibility between active material and fiber material. It is normally desirable that the two materials be mutually insoluble and that the fiber material be impermeable to the active material. This contributes to shelf life and assures that release is confined to the evaporation-diffusion-convection model.
2. Fiber material must have acceptable processing characteristics so that it can be spun, sealed, and cut to give clean unobstructed end openings.
3. Where food crop applications are involved, the fiber material must be acceptable for food contact use.
4. Chemically or biologically degradable fiber materials are preferred where dispensers are not recoverable in normal use patterns.

Generally, crystallizable polymers make the best hollow fiber evaporators. Most development work to date has involved either Celcon® (a poly-oxymethylene-co-oxyethylene copolymer made by Celanese Plastics Company, Chatham, N.J.) or poly(ethylene terephthalate). Polyolefins such as high density polyethylene, polypropylene, and poly (butene-1) have also been investigated for certain highly specialized applications with promising results. Virtually any polymer which can be spun into hollow fibers having a moderate degree of dimensional integrity and mechanical strength is potentially useful if other criteria bearing on a specific application can be met.

Other factors which have a bearing on hollow fiber controlled release formulation design are those dictated by the nature and economics of end use. Dose rate and longevity for an insect pheromone dispenser, for example, are determined by a multiplicity of factors including life cycle of the insect, biological response, growth characteristics of the host crop or plant, cultural practices, climate, pest management strategies, and methods of application. Controlled release formulations usually are more costly to produce than conventional delivery systems, and hollow fiber formulations are probably the most costly of those now in use on a commercial scale. In order for hollow fiber formulations to be competitive on a cost performance basis, full advantage must be taken of the strongest features of hollow fiber delivery systems which are (1) high efficiency of utilization of active material, (2) design flexibility, (3) reliability under field conditions, and (4) high ratio of active material to dissemination equipment payload.

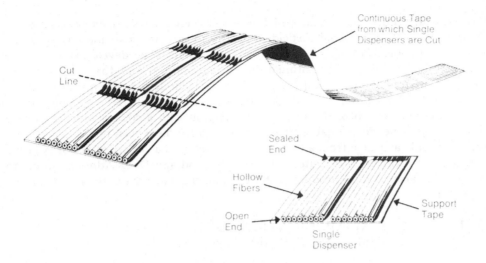

FIGURE 5. Tape form of hollow fiber vapor dispenser.

III. BASIC PRODUCT FORMS AND METHODS OF MANUFACTURE

A. Parallel Array or Tape-Type Dispensers

Hollow fiber controlled vapor release dispensers have been fashioned basically in two forms. In the first, a parallel array of fibers is fixed to an adhesive tape. After pressure filling the active material into the lumina, the fibers are sealed ultrasonically at regular intervals along the tape. Active material must be in a liquid form for the filling operation. Release is activated by cutting the tape at a point adjacent to the seals, thereby opening the fiber ends. A graphical illustration of this dispenser form is shown in Figure 5. This kind of dispenser is useful for the exact positioning of vapor point sources. Example uses for the tape-type dispenser would include insect attractant dispensers for baiting traps, dispensing vapor action insecticides to packages or containers of stored goods susceptible to insect damage, and deodorizing enclosed spaces with air-freshening fragrances or odor-masking chemicals.

Experimenters have found the parallel array system especially useful in establishing dose-response relationships with insect attractants. Cuthbert and Peacock used parallel array dispensers to advantage in establishing the optimum ratio of 4-methyl-3-heptanol (1), multistriatin (2), and α cubebene (3) in multilure, the aggregation pheromone of the smaller elm bark beetle.[8]

B. Chopped Fiber Dispensers

A second form of hollow fiber vapor dispenser is made by sealing individual fibers at regular intervals and then chopping at a predetermined distance from the seal. This chopped fiber form is designed primarily for broadcasting vapor point sources over large areas. Charging chopped fibers with an active material is accomplished by immersing the fibers in the material, drawing a vacuum to remove air and other gases from the lumen, and then releasing the vacuum. Again, the active material must be in a liquid form at the time of filling. Chopped fibers charged with active material are stored in hermetically sealed containers. Release is activated when the container is opened. Example applications for chopped fiber dispensers include broadcast dissemination of insect sex attractant pheromones for mating disruption of agricultural or forest insect pests (mating disruption with insect sex pheromones is discussed below), insecticide dispensing, and possibly soil fumigation.

Broadcast dissemination of chopped fibers was itself a challenging product development problem. Conventional pesticide dusting and spraying machinery is not suited to hollow fiber formulations, so special equipment had to be developed to make it possible to apply hollow fiber formulations in agricultural and forestry situations. Both ground and aerial dispensing equipment have been developed employing design concepts disclosed by Kitterman.[9] Applications of chopped fibers in agricultural and forestry situations are normally made in conjunction with a sticker which facilitates cling to foliage. The most suitable stickers found so far have been polybutenes which are commercially available from a number of companies in a variety of viscosity grades and degrees of tackiness. Polybutenes have U.S. Food and Drug Administration approval for use around food products, which makes them attractive for use in agricultural situations.

IV. PRACTICAL APPLICATIONS OF HOLLOW FIBERS FOR DISPENSING INSECT PHEROMONES

A. Insect Pheromone Technology and Applications in Pest Management

Insect pheromones are volatile organic compounds produced by insects for purposes of communication through their highly developed olfactory sensory systems.[10-12] These highly specific chemical cues influence insect behavior in a variety of ways. Pheromones are known to function as sexual attractants, stimulants, alarm signals, trail markers, aggregating signals, oviposition deterrents, and regulators of behavior in social insect colonies. Sex attractant pheromones have been especially interesting to entomologists, since they raise the possibility of controlling pest insects by interdicting sexual communication which leads to mating and proliferation of damaging populations.[13,14] Pheromones generally are relatively innocuous chemicals possessing a remarkably high order of biological activity. If population suppression can be accomplished with pheromones, it is reasonable to expect that they could become a more ecologically tolerable alternative to conventional insecticides for pest insect management. Indeed environmental considerations have motivated much of the research on insect pheromones.

Three basic strategies have been developed for the use of pheromones in insect pest management. These are (1) monitoring and survey trapping, (2) mass trapping, and (3) mating communication disruption.[14] Insect trapping with pheromones has proven to be a valuable tool in pest management programs. Traps enable pest management personnel to detect and locate infestations, time the application of control measures, and monitor the effectiveness of such measures. Mass trapping with pheromones is an approach to population suppression which involves physically removing insects from the environment in numbers sufficient to give an acceptable level of control. It amounts to a very sophisticated refinement of the flypaper concept. Mating communication disruption is an approach to suppressing pest insect populations which involves permeating the atmosphere over large areas with sex pheromone vapor in order to interfere with the intersexual communication process. By thus subverting the mating and reproduction process, population suppression can be achieved without resort to lethal agents which often produce detrimental side effects on local ecosystems. At the present time, it appears that mating communication disruption will become the most important use for insect pheromones, especially for suppression of lepidopterous pest species. One product based on this concept, Gossyplure H.F.® for suppression of pink bollworm on cotton, is destined for commercial introduction in 1978, and others of a similar nature can be expected to follow.[15,16] Mass trapping is somewhat limited in its potential because of the labor intensive nature of installing and maintaining large numbers of traps. For some insects, mass trapping seems to hold promise and continues to receive investigative and development effort. Survey and monitoring with pheromone traps is already a widely accepted commercial practice and is expected to grow in importance as the technology of integrated pest management reaches increasingly higher levels of sophistication.

Controlled release technology plays an important role in all use strategies for insect pheromones for two basic reasons. First, pheromones are typically very expensive materials to synthesize. Those which are commercially available may cost anywhere from several hundred to several thousand dollars per pound. Second, pheromones characteristically display a high order of biological activity and are therefore normally used in extremely minute quantities as compared, say, to conventional chemical insecticides. Application rates for mating communication disruption of lepidopterous species, for example, are on the order of grams per acre-season. Pheromone trapping lures require very precise metering of pheromone vapor release at extremely low dose rates. It is not unusual for the optimum dose rate of pheromone trap bait to be on the order of a few micrograms per day. If dose rate varies by much from the optimum for an attractant dispenser, reliability of the trapping information can suffer, a matter of no small consequence if such information is used as a basis for decision making. These factors dictate a controlled release vehicle for pheromone dispensing which is precise, reliable, and efficient. Without an adequate controlled release system, it would be difficult if not impossible to make pheromones practical and economical for use on a commercial scale.

B. Hollow Fiber Pheromone Formulations for Suppression of Agricultural Insect Pests
1. Pink Bollworm

Pink bollworm, *Pectinophora gossypiella*, is said to be one of the world's most serious cotton insect pests.[17] Unless controlled, this pest is capable of destroying 20 to 40% of a cotton crop and under some circumstances, may destroy a crop completely. Pink bollworm infestations occur in practically every cotton-growing country of the world. In the U.S., pink bollworm is present in Oklahoma, Texas, New Mexico, Nevada, Arizona, and California. In Oklahoma, Texas, and New Mexico, control of pink bollworm is maintained through cultural practices, but in the infested areas of Arizona and southern California, control is generally achieved by heavy use of insecticides. The U.S. pink bollworm infested area where insecticides are commonly used extends from central Arizona westward to the Colorado River basin and includes the Imperial, Palo Verde, and Coachella Valleys in southern California. Cotton production in this region generally totals 500 to 600,000 acres annually. The San Joaquin Valley in southern California is reported to have an established and growing native population of pink bollworm, but so far San Joaquin growers have not had to use insecticides for this pest. Since the San Joaquin Valley represents over one million acres of cotton production annually, this endemic but growing population is viewed as alarming economic and ecological threat by local cotton producers.

The female pink bollworm moth lays eggs under the calyx of immature cotton bolls or flower buds (squares). Upon hatching, larvae enter the boll or flower where they feed on seed or proteinaceous plant tissue. Just prior to pupation, the larvae exit the bolls, leaving a hole which exposes lint to fungal attack. Damage to both seed and lint represents a significant economic loss, and fungal attack often leaves seed contaminated with aflatoxin, adversely effecting seed marketability for human food use.

Contemporary control practices for pink bollworm are presently limited to insecticide applications and cultural practices. Organophosphate insecticides such as monocrotophos, guthion, and methylparathion are normally employed against pink boll-worm. The use of such materials initiated early in the season, anywhere from mid-June to mid-July, decimates predator-parasite complexes inviting subsequent out-breaks of secondary pests such as lygus, cotton leaf perforator, and the *Heliothis* bud-worm-bollworm complex.[18] In the 1977 cotton season, outbreaks of tobacco budworm, triggered by early season sprays for pink bollworm, resulted in catastrophic yield losses all across the desert Southwest. The Imperial Valley in California was most severely stricken, and there it is expected that 1977 yields will be about 500 lb of lint per acre down from around 1200 lb per acre in prior years. Unless viable alternatives to early season insecticide applications for pink bollworm control are found, continued cotton production in this part of the U.S. may become economically infeasible.[19]

Efforts to suppress pink bollworm infestations and reduce the need for insecticides on growing cotton by applying the mating disruption concept were first reported in 1974.[20] In the 1972 and 1973 growing seasons, Shorey, Kaae, and Gaston attempted to suppress pink bollworm on cotton in the Coachella Valley of southern California using hexalure, Z-11-hexadecadienyl acetate, a sex pheromone mimic, as the mating communication disruptant and experienced promising results.[21,22] In the 1974 season, Shorey and his colleagues treated the entire 4000 acres of cotton in Coachella Valley, using gossyplure as the disruptant, again with promising results.[23] Gossyplure is a 1:1 mixture of the Z,E and Z,Z isomers of 7,11-hexadecadienyl acetate, which has been identified as the true male sex attractant pheromone produced by female pink boll-worm moths.[24,25] In 1976 the Shorey-Gaston group turned to the use of Celcon® hol-low fiber gossyplure dispensers which were fashioned into hoops for exact placement and retention on cotton foliage. The 1976 experiments, conducted on small test plots, clearly established that disruption of pest insect pheromone communication can be a useful component of pest management strategy for crop protection. Pheromone treat-ment was shown to give control results quite comparable to conventional treatment practices based solely on insecticides (see Figure 6). These results translated into a ninefold reduction in insecticide usage on pheromone-treated fields as compared to conventional practice fields (see Figure 7).[26]

The first attempt to apply the mating disruption concept on a commercial basis was made in 1976 using Gossyplure H.F.®, a hollow fiber formulation of gossyplure pro-duced by Conrel, An Albany International Company.[15,16] This effort was pursued un-der an experimental use permit granted by the U.S. Environmental Protection Agency and was carried out on test sites located in central Arizona and the Imperial Valley in southern California. Test areas totaled approximately 2900 acres and numbered 64 individual fields, 57 in Arizona and 7 in California.

Gossyplure H.F.® used in the 1976 season was comprised of gossyplure diluted with hexane and contained in 200 μm I.D. Celcon® hollow fibers 1.75 cm in length and having a seal located at random distances from the open fiber ends. Hexane dilution served to adjust the length of the column of active pheromone in the field, which in turn governed release life. Early in the season, percentages of active pheromone in the formulation of up to 12.6% by weight were used. For most of the season, a single formulation containing 6.8% active pheromone was employed. This formulation cor-responded to a field life of approximately 14 days at 100°F (37.8°C).

Test areas were established in four locations: Tonopah, Rainbow Valley, and Buck-eye, Arizona and Imperial Valley, California. The Tonopah, Arizona area was com-prised of 24 cotton fields totalling 988 acres of contiguous properties. The Tonopah test fields were partially isolated by desert on three sides with only a road separating the nearest neighbor to the south. Next nearest neighbors in the Tonopah Valley were approximately two miles away to the southwest. Check fields totalling 316 acres were

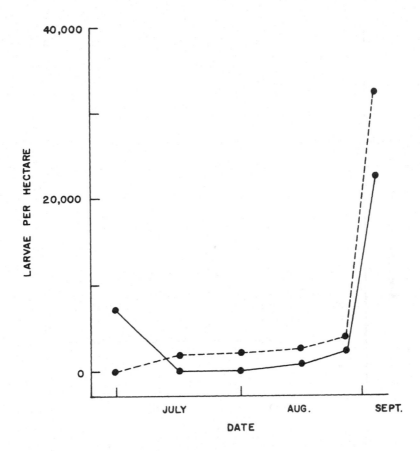

FIGURE 6.   The mean number of pink bollworm larvae per hectare in three pheromone-treated fields (solid line) and ten grower practice fields (dashed line). (From Gaston, L. K., Kaae, R. S., Shorey, H. H., and Sellers, D., *Science,* 196, 904, 1977. With permission.)

located on the adjacent southern cotton-producing properties and the more distant southwestern Tonopah properties. The Buckeye test area was comprised of six cotton fields totalling 131 acres, with an additional 110 acres serving as check fields. Experimental and check fields were all contiguous comprising a 241-acre block isolated on all sides by desert. The Rainbow Valley test area was comprised of 27 cotton fields totalling 1334 acres and 269 acres of check fields. Rainbow Valley experimental fields were interspersed with fields under conventional insect control practices and could not be considered as well isolated. In Imperial Valley, California, there were seven experimental fields totalling 469 acres and two check fields totalling 170 acres. Imperial Valley test fields were not isolated from cotton producing neighbors.

Gossyplure H. F.® was applied in broadcast fashion early in the season by ground equipment and later by air. Equipment was designed to apply a sticker material, polybutene, to fibers as they were dispersed to promote cling to the cotton foliage. Rates of application were somewhat imprecise owing to difficulties in maintaining calibration on the experimental fiber dispensing equipment, but were generally in the range of 1 to 3 of active material per acre per application. The average number of applications was seven per field at intervals of 10 to 20 days with 10-day intervals during the warmest part of the season in late July and August. The number of pheromone point sources per acre at the 1 to 3- g rate of application was in the range of 3 to 9,000. Disruption

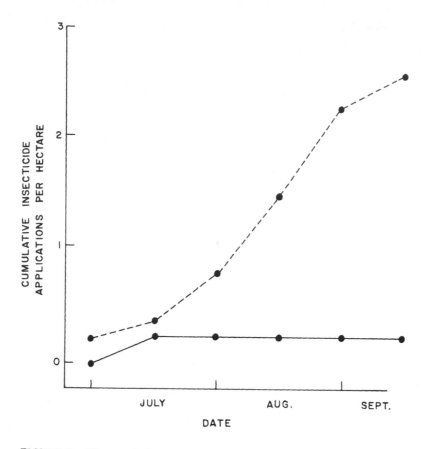

FIGURE 7.   The cumulative mean number of insecticide applications per hectare
for three pheromone-treated fields (solid line) and ten grower practice fields (dashed
line). (From Gaston, L. K., Kaae, R. S., Shorey, H. H., and Sellers, D., *Science*,
196, 904, 1977. With permission.)

was monitored on pheromone-treated fields by daily reading of pheromone-baited om-
nidirectional traps stationed at a density of one trap for every ten acres. When traps
began capturing male moths, a fresh application of Gossyplure H.F.® was made. Since
check fields were under treatment with insecticide for most of the season, percent disruption on treated vs. nonpheromone fields was not applied to gauge treatment efficacy,
although moth captures outside of treated areas were consistently much higher than
within pheromone-treated fields. On all experimental and check fields, sampling of susceptible bolls for larval infestation was conducted on a twice weekly basis. Each ten-
acre section was sampled by pulling 20 bolls. Percent boll infestation was calculated
from weekly averages of the two samplings.

In the Tonopah test area, larval infestation on the 24 pheromone-treated fields remained below 1% from mid-May when treatments commenced until mid-July when
seven of the test fields sustained infestations exceeding 10%. By contrast, Tonopah
check fields experienced bloom infestations as high as 50% in mid-June and were
placed on 5- to 7-day insecticide spray schedules. Pheromone treatments in Tonopah
apparently postponed the onset of significant pink bollworm infestation in that area
by about 1 month or one generation. Pheromone treatment was abandoned at this
point, and all of the Tonopah test fields placed on insecticide spray schedules. A net
reduction in insecticide use on pheromone vs nonpheromone fields of 30% was experienced for the 1976 season. On the six pheromone-treated fields in the Buckeye test

area, larval infestation remained below 1% for the entire season. This resulted in a 33% net reduction in insecticide use for pink bollworm control on pheromone vs. nonpheromone fields. In the Rainbow Valley test area, only 1 out of 27 pheromone-treated fields sustained a larval infestation exceeding 1%. The one problem field sustained a 30% larval infestation in the last week of August. Three different growers were involved in the Rainbow Valley pheromone experiments. The first had nine test fields totalling 504 acres and experienced a 47% reduction in insecticide use when compared to check fields. The second had 16 fields totalling 593 acres and experienced a 61% reduction in insecticide use. The third grower had two fields under pheromone treatment totalling 205 acres and showed a 68% reduction in insecticide use. In Imperial Valley six of the seven pheromone-treated fields sustained less than 1% larval infestation throughout the season. The one problem field in Imperial Valley was an eight-acre plot surrounded by about 400 acres of nonpheromone cotton. This small plot held at a low level of infestation until mid-August when damage jumped to 20% larvae infested susceptible bolls. An overall reduction of about 50% in insecticide applications in Imperial Valley pheromone fields was realized.

In the 1977 growing season, test acreage for Gossyplure H.F.® was increased to approximately 23,000 acres. Nine test areas were established, seven in Arizona and two in California. The Gossyplure H.F.® formulation used in 1977 was modified slightly by shortening the fiber from 1.75 to 1.5 cm, resulting in a label percentage of active ingredient of 7.6%. Application machinery and methodology was improved over 1976, but otherwise basic use patterns were the same in 1977 as in 1976. Treatment programs in 1977 were under the management of commercial agricultural chemical field men and consulting entomologists, most of whom were being introduced to pheromone technology for the first time. The basic strategy was to employ Gossyplure H. F.® for early season suppression of pink bollworm populations, avoiding the use of conventional insecticides as much as possible, but using them as necessary to control any outbreaks of secondary pests.

Preliminary use data for Gossyplure H.F.® from the 1977 season were available at the time of this writing for approximately 25% of the treated acreage. These data are summarized in Table 2 to show the reduction in insecticide usage realized with mating communication disruption treatments as compared to conventional practice treatment methods. The relative costs to growers of pheromone vs. conventional practice methods are also shown.

The data in Table 2 show that use of Gossyplure H. F.® for early season suppression of pink bollworm population resulted in 50 to 80% reductions in the use of chemical insecticides for protection of the cotton crop from insect damage. The data from the Cibola test area near Blythe, California were broken down further into four categories of treatment regimes: (1) fields treated by conventional practice (insecticide only), (2) fields receiving only pheromone treatment, (3) fields receiving treatments with both pheromone and insecticide, and (4) fields receiving treatments with pheromone, *Bacillus thuringiensis* (B.T), and insecticide. This second analysis is shown in Table 3.

In Table 4 early harvest results from the Blythe, California test area are presented, showing rather dramatically the potential benefits of pheromone-assisted pest management schemes in terms of crop yields improvements. The relatively poor yields on untreated check and conventional practice acreage are attributable largely to insect damage caused by pink bollworm and tobacco budworm (*Heliothis virescens*).

Results thus far with Gossyplure H. F.® give every reason to believe that insect pheromones, formulated in a suitable controlled release vehicle, can be an effective and economical adjunct to cotton pest management programs. The 1977 experimental

TABLE 2

Comparison of Number of Applications, Weight of Active Ingredients, and Acre-Season Costs for Conventional Practice (Insecticide Only) vs. Gossyplure H. F.® Insect Pest Management Programs, 1977 Season

Aquila, Arizona, Test Area

	Conventional Practice	Gossyplure H.F. Program	% Reduction (Increase)[a]
No. of acres analyzed	498.6	1452	—
Average no. of insecticide acre applications	7.2	4.81	33.2%
Average no. of acre applications for insect control	7.2	8.9	(23.5%)
Average active ingredient of insecticide applied per acre-season	7.8 lb	1.6 lb	79.3%
Average Gossyplure H. F.® active ingredient used per acre-season	—	4.8 g	—
Average cost per acre-season for insect control[b]	$46.21	$58.45	(26.5%)

Blythe, California Test Area (Cibola)[c]

	Conventional Practice	Gossyplure H.F. Program	% Reduction (Increase)[a]
No. of acres analyzed	658	2060	—
Average no. of insecticide acre applications	9.3	2.8	69.7%
Average no. of acre applications for insect control	9.3	8.4	9.6%
Average active ingredient of insecticide applied per acre-season	18.96 lb	3.93 lb	79%
Average Gossyplure H.F.® active ingredient used per acre-seaso	—	5.36 g	—
Average cost per acre-season for insect control[b]	$80.17	$58.75	27%

Yuma, Arizona Test Area

	Conventional Practice	Gossyplure H.F. Program	% Reduction (Increase)[a]
No. of acres analyzed	951	420	—
Average no. of insecticide acre applications	9.7	3.3	66.3%
Average no. of acre applications for insect control	11.5	6.9	40%
Average active ingredient of insecticide applied per acre-season	15.8 lb	4.2 lb	73.2%
Average Gossyplure H.F.® active ingredient used per acre-season	—	4.1 g	—

TABLE 2 (continued)

Comparison of Number of Applications, Weight of Active Ingredients, and Acre-Season Costs for Conventional Practice (Insecticide Only) vs. Gossyplure H. F.® Insect Pest Management Programs, 1977 Season

Average cost per acre-season for insect control[b]	$75.50	$47.60	37%

Parker, Arizona Test Area

No. of acres analyzed	213	168	—
Average no. of insecticide acre applications	9.5	3.9	58.9%
Average no. of acre applications for insect control	9.7	10.1	(4.0%)
	9.7	10.1	(4.0%)
Average active ingredient of insecticide applied per acre-season	18.3 lb	7.6 lb	58.2%
Average Gossyplure H.F.® active ingredient used per acre-season		6.8 g	
Average cost per acre-season for insect control[b]	$87.50	$79.43	9.2%

Gila Bend, Arizona Test Area[d]

No. of acres analyzed	1,632	1,433	—
Average no. of insecticide acre applications	9.5	7.0	26.5%
Average no. of acre applications for insect control	9.5	15.1	(28.5%)
Average active ingredient of insecticide applied per acre-season	25.6 lb	11.5 lb	55.2%
Average Gossyplure H.F.® active ingredient used per acre-season	—	8.5 g	—
Average cost per acre-season for insect control[b]	$117.02	$99.69	14.8%

Imperial Valley, California Test Area

No. of acres analyzed	850	429	—
Average no. of insecticide acre applications	10.2	4.0	60.5%
Average no. of acre applications for insect control	10.2	9.4	7.9%
Average active ingredient of insecticide applied per acre-season	13.1 lb	6.1 lb	54.0%

TABLE 2 (continued)

Comparison of Number of Applications, Weight of Active Ingredients, and Acre-Season Costs for Conventional Practice (Insecticide Only) vs. Gossyplure H. F.® Insect Pest Management Programs, 1977 Season

Average Gossyplure H.F.® active ingredient used per acre-season	—	5.2 g	—
Average cost per acre-season for insect control[b]	$114.16	$78.58	31.0%

[a] Reduction in insecticide usage on Gossyplure H. F.®-treated acreage as compared to conventional practice acreage indicates that pheromone suppression gave crop protection results comparable to that achieved solely with insecticides and is a measure of the reduction in chemical stress on the local ecosystem.

[b] Cost is for insect control materials only. It does not include application expenses or pest management consulting fees. Costs were calculated from published price lists for the chemicals used.

[c] The Cibola area in Blythe was well isolated from external sources of pink bollworm infestation except for a single 100-acre field in the pheromone treatment block which was not treated with any insect control chemicals at all. The one untreated field sustained heavy pink bollworm damage which serves to indicate the pink bollworm pressure in this valley.

[d] In this test area, the conventional practice and Gossyplure H. F.® treatment blocks are almost identical as to area size, degree of isolation by surrounding desert, geographical circumstance, insect pressures, and past history of pink bollworm infestation.

TABLE 3

Summary of 1977 Gossyplure H. F.® Treatment Results in Blythe, California (Cibola Valley) by Treatment Regimen[a,b]

	Conventional practice	Pheromone plus insecticide	Pheromone plus B.T. and insecticide	Phero-mone only
Average no. of applications	9.3	7.9	8.6	5.7
Average lb per acre Insecticide A. I.[c]	18.96	12.07	3.94	0
Average cost per acre Season	$80.17	$61.99	$67.65	$35.73
Average % reduction in hard chemicals	—	36%	79%	100%
Average % reduction in cost	—	23%	16%	55%
Total acres analyzed	685	254	1277	529
Percent of total	25%	9%	47%	19%

[a] 74% of Gossyplure H. F.® acreage had insecticide applications vs. 100% of conventional practice fields

[b] 62% of Gossyplure H. F.® acreage had B. T. applications vs. 27% of conventional practice acreage.

[c] Average amount of Gossyplure H. F.® active ingredient applied was 5.4 g per acre-season.

TABLE 4

Early Yield Information on 1977 Gossyplure H. F.®-Treated Cotton Acreage in the Palo Verde Valley (Blythe, California — Cibola)[a]

			Yield per acre	
Source	Acres picked	No. of[c] picks	Lbs	Bales
Untreated[b]	100	3	569	1.07
Test Farm A	185	3	1012	1.94
Test Farm B	509	3	1005	1.92
Test Farm C	120	1	1210	2.37
Test Farm D	36	2	1102	2.14
Conventional practice	—	3	—	1.25—1.50 (estimated)[d]

[a] Harvest results available as of 12/2/77.
[b] The untreated check farm used no chemicals at all. In late August, a random sampling of the check field revealed 83 to 88% of susceptible bolls infested with pink bollworm.
[c] Less than three picks indicates harvest was not yet completed.
[d] Ginners estimate for all of the Palo Verde Valley, based on yields reported through 12/2/77. Palo Verde Valley had approximately 43,000 acres in cotton production in 1977.

program with Gossyplure H.F.® was the largest effort recorded to date utilizing the mating communication disruption concept for crop protection. The outcome of this effort seems almost certain to establish this use of pheromones as a commercial reality.

2. Grape Berry Moth

Grape berry moth (*Paralobesia viteana*) is the most important fruit-infesting pest of grapes in growing areas east of the Mississippi River. The female-produced male sex attractant pheromone of the grape berry moth is (Z)-9-dodecenyl acetate.[27] In 1975, Taschenberg and Roelofs demonstrated orientation disruption of grape berry moth with pheromone formulated in both microcapsules and hollow fibers using the atmospheric permeation method.[28] Hollow fiber formulations used in the tape form and applied one time at rates of 1.0 and 1.5 g per acre were effective in sustaining disruption for over 2 months. Microencapsulated pheromone was applied at 5 to 7 day intervals over the season for a total of 200 g per acre. Lack of persistence of microencapsulated material made this type of formulation unattractive economically when compared to the performance of hollow fiber formulations.

In 1976, the grape berry moth disruption experiments were repeated with both chopped fiber and tape-type hollow fiber pheromone formulations.[29] Single pheromone treatments on two test plots with tape-type dispensers at rates of 0.8 mg/hr per acre (50 fiber ends per vine) and one half that rate (50 fiber ends on alternate vines) produced 99.5 and 97.5% reduction in male orientation to pheromone traps, respectively, for a period of 105 days. Fruit damage was 0.4 and 0.6%, respectively, as compared to a check plot showing 1.7% damage. A third test plot was treated twice, on June 3 and July 17, with chopped fiber pheromone formulation providing a release rate of 0.8 mg/hr per acre. Male orientation disruption was 92.7% and fruit damage on the treated plot was 1.9 vs. 8.5% on the check plot. The orientation disruption results for both types of dispensing formulation are shown in Figures 8 and 9. While the crop protection results in these experiments are definitely encouraging, the level of observed fruit damage in check plots was too moderate to warrant any final conclusions concerning efficacy. The important implications in these results lie in the demonstrated capability of obtaining season-long disruption at rates of pheromone application that are economically feasible.

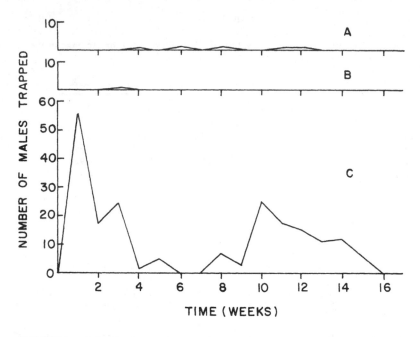

FIGURE 8.   Orientation disruption of grape berry moth. Male catches from 5/28/ 76 to 10/17/76 in four pheromone traps per plot. Treatments consisted of (A) 50 fibers per alternate vine (0.4 mg pheromone per hour per acre), (B) 50 fibers per vine (0.8 mg pheromone per hour per acre), and (C) Check plot. (From Taschenberg, E. F. and Roelofs, W. L., *Environ. Entomol.*, 5, 761, 1977. With permission.)

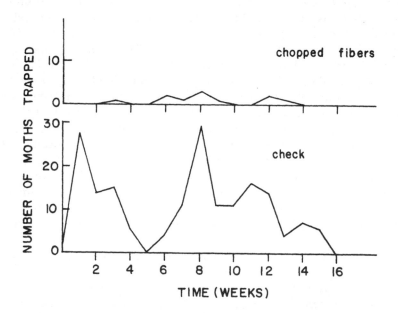

FIGURE 9.   Orientation disruption of grape berry moth. Male catches from 6/13/76 to 10/16/76 in four pheromone traps per plot. Plot treated with chopped fiber on 6/3/76 and 7/17/76 to provide pheromone release rate of 0.8 mg/hr per acre. (Bottom) Check plot. (From Taschenberg, E. F. and Roelofs, W. L., *Environ. Entomol.*, 5, 761, 1977. With permission)

3. Codling Moth

The codling moth (*Laspeyresia pomonella*) is a major world-wide pest attacking apples, pears, walnuts, and other tree fruits. The combined costs of crop damage and control measures owing to codling moth are estimated at 39 million dollars annually in the U.S. alone. The sex pheromone of codling moth has been identified as (*E,E*)-8,10-dodecadienol[30,31] and bears the trivial name codlure. A small test plot experiment with codlure formulated in hollow fibers was conducted in Michigan in 1976 by Carde and co-workers.[32] Hollow fibers releasing codlure at a rate of 8.5×10^{-4} mg/hr (21°C) were distributed in apple trees so that there were ten fibers at about 1700 different places in the 72-tree plot. The ten-fiber point sources were of the parallel array tape type. Total release was 0.8 mg/hr per acre. Female-baited traps attracted no codling moth males in the disruption test plot, compared to 383 males trapped in an untreated check plot. Pheromone-baited traps in the test plot caught no males from May 21 to August 7, compared to 284 males caught in pheromone-baited traps on the check plot. Up to August 13, pheromone traps caught five males on the test plot, while traps on the check plot had a cumulative catch of 787 males. There was also disruption of the mating of tethered virgin female moths, although mating of male and female pairs in small cages was not affected. These data are taken as an indication that disruption of long-range orientation of codling moth can be an effective control measure where population densities are low enough to prevent many chance sexual encounters.

In the 1977 season, Moffitt and co-workers conducted codling moth disruption experiments with chopped fiber codlure formulations in a pear orchard near Medford, Oregon.[33] Chopped hollow Celcon fibers loaded with codlure were distributed at a rate of approximately 35,400 per acre to provide a nominal release rate of 5.3 mg/hr per acre. Applications were made by helicopter, using polybutene sticker on each of two one-acre test plots with 70 trees each. Within the 6.4-acre orchard, a standard insecticide program for codling moth control was applied to a plot of the same size, and a block of 20 trees served as untreated control. Pheromone applications were made April 21, May 25, and July 26. Harvest of the Bartlett variety was August 19 and of the Bosc variety on September 12.

Efficacy determination in the Moffitt disruption experiment was based upon response of both native and laboratory-reared male moths to either pheromone or female-baited traps. Orientation disruption as measured by trap catches was inconclusive because the pheromone treatment apparently affected male response throughout the relatively small test area. On the early maturing Bartlett variety, the infestation level at harvest was 0.05%, while on the later maturing Boscs it was 0.4%. In the untreated control plot, an infestation level of 4 to 9% was recorded.

While the results of these efforts to suppress codling moth infestations by mating communication disruption do not warrant a firm conclusion about efficacy, they clearly lend encouragement to future work. A definite efficacy statement will probably require at least another season of testing preferably on larger treated areas. How codlure may eventually fit into integrated programs for orchard pest management remains to be seen.

C. Forestry Pests

1. Spruce Budworm

The eastern spruce budworm (*Choristoneura fumiferana*) is the major pest of white spruce and balsam fir in eastern North America. It is a native species whose history is characterized by long periods of endemism interspersed with spectacular epidemic periods. Historically, intervals between epidemics have varied from 21 to 100 years, and recent history indicates a trend to more frequent outbreaks. Epidemic periods usually last about 10 years, if left unchecked, and tree mortality can be extensive. The present epidemic in eastern North America is believed to be the worst ever and has resulted in heavy defoliation over several years throughout more than 100 million acres of forest. This in turn has led to massive insecticide spray programs over millions of acres annually in an effort to spare timber and pulpwood stands and recreational forest. Fears over the environmental consequences of such massive spray programs have stimulated research on alternative forest protection methods, including potential species specific control agents such as sex pheromone.

The natural sex attractant of the eastern spruce budworm has been identified as a 97:3 blend of the (*E*) and (*Z*) isomers of 11-tetradecenal.[34] This blend has been assigned the trivial name fulure. In 1977, Sanders and co-workers attempted to demonstrate population suppression of eastern spruce budworm, using hollow fiber-formulated fulure on two test plots, one of moderate and one of low population density in the Province of Ontario.[35] While it is generally recognized that the mating communication disruption concept is most likely to succeed in low density populations, for logistical and other reasons a moderately high population test area was deemed desirable.

In the first test near Sault Ste. Marie, Ontario a 600-acre treated plot of white spruce was established with a nearby check plot of some 30 acres. Populations on this site in 1977 were moderate to high initially, while populations and host tree numbers in the surrounding area were low so that gravid female invasion was improbable. Fulure-loaded chopped fibers were applied aerially in combination with polybutene sticker over a period of 8 days. Weather conditions and mechanical problems prolonged the application period. The chopped fiber formulation was designed to give a fulure delivery rate of 12 mg/hr per acre for 5 weeks at an application rate of 73,000 fibers per acre. A total of 24,000 fibers per acre was actually released, but drop sheets verified a rate of only 13 to 14,000 for a calculated air permeation rate of 2 mg/hr per acre.

The disruption experiment was monitored by (1) prespray larval and pupal sampling, (2) postspray trapping in test and check plots with virgin female baited traps, fulure baited traps, and double ended lobster pot traps containing single untethered virgin female budworm moths, (3) postspray egg sampling, (4) postspray sampling of overwintering larval populations, and (5) observations from 50-ft scaffolds erected in test and check plots. Test results can be summarized as follows:

1. Prespray sampling revealed no significant differences between test and check plots. They did reveal an unusual collapse in populations (from 32 larvae to 0.4 pupae per 45-cm branch tip), effectively reducing population to a low density.
2. Trapping results showed (a) cumulative catches per virgin female-baited trap in the test plot averaged 2.2 vs. 394.4 in the control plot (99.4% disruption), (b) cumulative catches in fulure-baited traps in the test plot averaged 22.8 per trap vs. 840.6 in the control (97.3% disruption), and (c) in the live traps, out of 110 opportunities for mating, 35 females were found with males present in the trap, 25 of which showed evidence of successful mating (spermatophores found), while in the test plot, only two females were found with males and neither had mated successfully.
3. Because of the application delays, egg samples were taken on June 30 to detect any evidence of oviposition prior to application and on July 27 after the moth flight. The June 30 sampling yielded 0.36 ± 0.13 egg masses per branch tip in the control plot as compared with 1.16 ± 0.35 in the test plot, a considerably higher incidence of oviposition in the test plot prior to application. On July 27,

however, the check plot showed a significantly lower count of 0.71 ± 0.11, while the test plot count was 0.97 ± 0.12, almost the same as earlier.

4. Preliminary larval counts showed higher overwintering populations in the check plot than in the test plot, which contradicts the egg counts. Interpretation is difficult, however, in light of the unanticipated population collapse. Also, budworm larvae were outnumbered by other species of larvae, so positive identification of the extremely small larvae further complicated data collection and analysis.

The result of the high population density experiment with fulure was an apparent reduction in the density of the succeeding generation although this result is obscured by the preapplication oviposition on both plots and the population collapse, which made counts very low. The egg counts also are not corroborated by counts of overwintering larvae. While efficacy in terms of reduced population is questionable, orientation disruption of adult moths is clearly not, which testifies well for performance of hollow fiber fulure formulation in terms of meeting design specifications for longevity.

In the low-density population experiment conducted close to Black Sturgeon Lake near Thunder Bay, Ontario, two 25-acre test plots were established separated by about 2 km. A control plot was some 15 km away from the treated area. Aerial applications were made at rates of 73,000 fibers per acre (12 mg/hr per acre) on one plot and 7300 fibers per acre (1.2 mg/hr peracre theoretical) on the other. Deposit trays in the higher rate plot did not verify the apllication rate, and the discrepancy could not be resolved. Trapping data showed 99.4% orientation disruption on the high rate plot and 97.3% on the low. Mating of live virgin females was also denied as in the high-density experiment. In spite of the profound disruption results, postspray larval counts showed no measurable effect from the treatment. Explanation of the apparent failure may lie in the lateness of the application, i.e., mating and oviposition had already occurred prior to treatment, or the test plots could have been invaded by gravid females from outside the test area. Future research will have to explore these questions since the overall results are otherwise promising, and the delivery vehicle performs to specifications on the basis of trapping results.

2. Tussock Moth

The Douglas fir tussock moth (*Orgyia pseudotsugata*) is a severe defoliator of fir forests in western North America. This forest pest is given to population explosions which result in severe damage for 1 or 2 years before the population subsides. These sporadic outbreaks cause many millions of dollars in timber losses, place forests in jeopardy of further fire losses, ruin wildlife habitat, and spoil forest which serves as human recreational area. They are also something of a health hazard to humans who work in forests (loggers, for example) because hairs of the moths produce allergic responses such as welts, eczema, rashes, and breathing difficulties.[36] Historically, tussock moth epidemics have been quelled with DDT, but the possibility of food chain contamination and considerable public controversy has essentially eliminated DDT as an available remedy. In recent years, research has been directed to developing ecologically acceptable alternative control agents such as viral or bacterial pathogens and pheromones.

The sex pheromone of the Douglas fir tussock moth has been identified as (Z)-6-heneicosen-11-one by Smith et al.[37]

$$CH_3(CH_2)_9 CO(CH_2)_3 \diagdown \qquad \diagup (CH_2)_4 CH_3$$
$$C = C$$
$$H \diagup \qquad \diagdown H$$

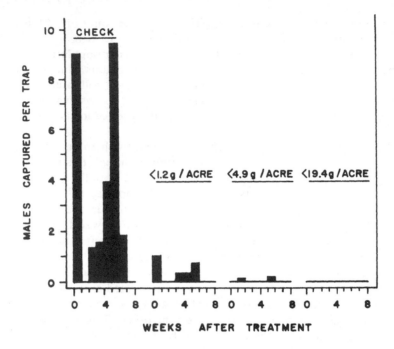

FIGURE 10.   Douglas fir tussock moth male moths captured per trap baited
with live females in check plots in the presence of three dose rates of pheromone applied in chopped fibers.[40]

Sower and Daterman used this pheromone successfully for mating communication disruption of Douglas fir tussock moth,[38] and Grant more recently found that it also
caused disruption of white marked tussock moth (*Orgyia leucostigma*) and pine tussock moth (*Dasychira plaginata*).[39]

In 1977, Daterman and Sower investigated Douglas fir tussock moth mating disruption with (Z)-6-heneicosen-11-one under both outbreak (high) and endemic (low) population density circumstances, using chopped fiber formulations.[40] In the low population experiments, nine test plots, five acres each, were established along with four
check plots near Fort Klamath, Oregon. Test plots were treated aerially at nominal
rates of 4000, 15,000, and 61,000 fibers per acre. Three plots were treated at each rate.
These rates correspond to 1.2, 4.9, and 19 g of active pheromone per acre. Actual
treatment rates were lower than these figures because many fibers were released before
the aircraft entered the test plot and after passing the plot. Mating disruption was
gauged by trapping in the test and check plots with virgin female baited traps and by
taking counts of egg masses produced by wild females.

In the check plots, male moth trap captures were 3.4 per week per trap, while on
the test plots captures were 0.1, 0.06, and 0.0 males per trap-week at the respective
dose rates (see Figure 10). Traps placed at levels of 5 and 60 ft above ground gave
similar capture readings. Egg mass counts (Figure 11) showed reductions in egg laying
of 73, 85, and 93% relative to untreated plots at the respective dose rates. These results
indicate that atmospheric permeation with (Z)-6-heneicosen-11-one will substantially
reduce the incidence of mating of Douglas fir tussock moth under endemic population
circumstances.

A similar test was conducted on a single ten-acre plot in Medio Dia Canyon, New
Mexico where a high population of Douglas fir tussock moth had already resulted in
partial or complete defoliation of fir trees. Pheromone formulated in hollow fibers

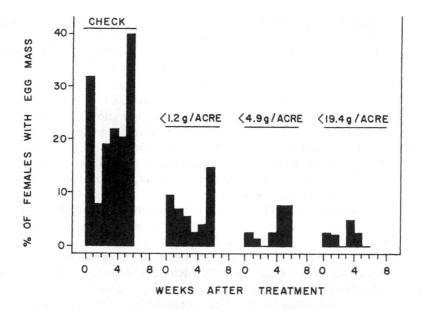

FIGURE 11. Production of Douglas fir tussock moth egg masses in check plots and in the presence of three dose rates of pheromone.[40]

was applied aerially at a rate of 190 g per acre (19 g per acre active ingredient). Females emerging from tagged cocoons deposited eggs at a rate of 0.92 ± 0.10 egg masses per female in the untreated check plot, but only 0.21 ± 0.12 in the treated plot, a mating reduction of 77%. While a substantial reduction in mating incidence was observed in the high population density experiment, pheromone treatment was less effective than comparable treatments of a lower population density. These results were attributed to the differences in population density, although the apparent population effect lacks statistical confirmation. Based on these results there is every reason to believe that mating communication disruption with (Z)-6-heneicosen-11-one may one day afford a viable and ecologically innocuous means of suppressing Douglas fir tussock moth infestations and avoiding the severe forest damage produced by outbreaks.

V. LIMITATIONS OF HOLLOW FIBERS AS A CONTROLLED RELEASE VEHICLE

A. Technical and Economic Factors Affecting Commercial Feasibility

In the few short years since the hollow fiber concept of controlled vapor release first came into being, considerable progress has been made toward commercial applications, particularly with the use of insect pheromones in pest management. Like every other controlled release vehicle, however, hollow fibers have distinct technical and economic limitations which govern their applicability in a given situation. On the technical side, hollow fibers are limited by (1) an inherently low release capability in terms of mass released per fiber, (2) a high weight ratio of delivery vehicle to active material, and (3) the requirement for specialized machinery where controlled release dispensers are applied in broadcast fashion. In applications where a relatively high output of active vapor is necessary — certain fragrances for example — hollow fiber systems have proven inadequate when compared to simpler methods such as cotton wicks or gel formulations. This can be attributed to the inherently large differences in evaporative surface areas. Where large surface areas are important, hollow fibers are comparatively limited. Hollow fiber formulations typically have a relatively high ratio of vehicle to active material, especially when they are compared to microencapsulation formulations. Where the vehicle mass represents a significant incremental portion of formulation cost and there are no offsetting advantages for hollow fibers such as more efficient utilization of active material, other kinds of formulation are often preferable. Where controlled release formulations are to be deployed in broadcast fashion over large areas, hollow fibers will require specialized dissemination equipment as contrasted with microencapsulated formulations which may be deployed with conventional spray application equipment used in agriculture and forestry. This limitation of hollow fiber formulations is becoming less significant as application machinery developments continue, but is still a definite bottleneck to rapid commercial exploitation of the technology in agriculture and forestry.

In the economic area, hollow fiber formulations are limited mainly by the relatively high unit cost for materials and processing. Where the economics of an end use are very sensitive to the unit cost of formulating, hollow fibers are often at a disadvantage over other less costly options. When the active material has a high unit value, however, hollow fiber formulations can be economically advantageous through better efficiency of utilization. This factor, perhaps more than any other, has made insect pheromones an attractive field for exploiting the technical potential of hollow fiber vapor release systems. Indeed the commercial feasibility of pheromone technology is strongly tied to the performance capabilities of a controlled release vehicle. Without an efficient delivery vehicle, the expense of synthetic pheromones would become a significant economic deterent to their large-scale use.

B. Environmental Factors and Regulatory Constraints

The potential environmental as well as economic benefits of controlled release technology, particularly in the field of pest control, are well documented.[41] In this regard, hollow fiber controlled vapor release devices add to a growing stable of controlled release vehicles and methods which promise to make pest management chemicals more efficient, less hazardous to handle, and less injurious to the environment. The environmental impact of controlled release products and the residues they leave is expected to be rather nominal, if even discernable, in most cases. While the ultimate environmental fate of most polymeric controlled release matrices has not been thoroughly studied, reasoned speculation leads to little expectation that vehicle residues will have any adverse consequences.[42]

Hollow fiber formulation studies have thus far focused almost exclusively on insect pheromones and the use of pheromones as alternatives to conventional pesticides in agriculture and forestry. Such studies have revealed an enormous potential for properly formulated pheromones in pest management schemes designed to protect crops or forest land while minimizing chemical stress on the environment. In this context, the net environmental effect of hollow fiber formulated pheromones used for pest management purposes is expected to be overwhelmingly favorable.

Controlled release pest control products are subject to regulation under the Federal Insecticide Fungicide and Rodenticide Act (FIFRA) of 1947 as amended by the Federal Environmental Pesticide Control Act (FEPCA) of 1972. No distinctions are made between controlled release and more conventional pesticide formulations under this law. Each new formulation must stand on its own merits with respect to meeting efficacy, safety, and environmental standards for registration, regardless of whether or not the active ingredient has received prior registrations. Insect pheromones or other types of attractant or repellent chemicals are considered pesticides by definition under FIFRA in spite of their nonlethal mode of action. Administrative policy within the U.S. Environmental Protection Agency, where FEPCA places regulatory authority for pesticides, makes no distinctions among pest control products as to the nature of their biological activity, specificity, or degree of hazard potential. Since the expense of registering new pest management products can be considerable, present policy often makes the registration process an economic disincentive. This is particularly true where highly specific products such as pheromones are involved because of their inherently limited marketability. Section 20(a) of FEPCA specifically directs the administrator of EPA to "develop biologically integrated alternatives for pest control". The law on the one hand seems to foster development of biologically rational products, while administration of the law on the other hand removes economic incentives to do so. It is unclear at the moment how this apparent contradiction will be resolved. Until some resolution is reached, the regulatory impact of FEPCA on product innovations such as controlled release pesticides and insect pheromones will be difficult to define.

Celcon hollow fibers and polybutene sticker have been granted exemptions from the requirement of a tolerance when used as inert materials with gossyplure, sex pheromone of the pink bollworm.[43] Gossyplure H.F.®, a hollow fiber controlled release formulation of the pink bollworm sex attractant, was granted registration by the U. S. Environmental Protection Agency on February 9, 1978. This is the first insect pheromone product to be registered for protection of a field crop (cotton).

VI. CONCLUSION

Spurred by needs and opportunities in pheromone technology, controlled vapor release from hollow fibers has moved from concept to commercial reality in less than 4 years. The utility of hollow fiber formulations for dispensing insect pheromones is now well established, promising an interesting future for both controlled release technology and insect pest management. The potential benefits of these two technologies to man and the environment are enormous. The extent of which benefits are realized will depend upon many imponderables not the least of which is their ability to measure up in the arena of commerce.

REFERENCES

1. **Brooks, T. W., Ashare, E., and Swenson, D. W.,** Hollow fibers as controlled vapor release devices, in *Textile and Paper Chemistry and Technology*, Arthur, J. C., Jr., Ed. ACS Symposium Series 49, American Chemical Society, Washington, D. C., 1977, 111.
2. **Ashare, E., Brooks, T. W., and Swenson, D. W.,** Controlled release from hollow fibers, in *Controlled Release Polymeric Formulations*, Paul, D. R. and Harris, F. W., Eds., ACS Symposium Series 33, American Chemical Society, Washington, D. C., 1976, 273.
3. **Coplan, M. J. and Brooks, T. W.,** Devices for Controlled Release of Vapors, U. S. Patent 4,017,030, April 12, 1977.
4. **Lonsdale, H.,** private communication, 1978.
5. **Bird, R. B., Stewart, W. E., and Lightfoot, E. N.,** *Transport Phenomena,* John Wiley & Sons, New York, 1960.
6. **Pippard, A. B.,** *The Elements of Classical Thermodynamics,* Cambridge University Press, London, 1960.
7. **Pearce, G. T., Gore, W. E., Silverstein, R. M., Peacock, J. W., Cuthbert, R. A., Lanier, G. N., and Simeone, J. B.,** Chemical attractants for the smaller European elm bark beetle, *Scolytus multistriatus* (Coleoptera: Scolytidae), *J. Chem. Ecol.,* 1, 115, 1975.
8. **Cuthbert, R. A. and Peacock, J. W.,** Response of *Scolytus multistriatus* to component mixtures and doses of the pheromone, multilure, *J. Chem. Ecol.,* 4, 375, 1978.

9. **Kitterman, R. L.,** Broadcast Dissemination of Trace Quantities of Biologically Active Chemicals, U.S. Patent 3,994,437, November 30, 1976.

10. **Birch, M. C., Ed.,** *Pheromones,* Elsevier, New York, 1974.

11. **Shorey, H. H.,** *Animal Communication by Pheromones,* Academic Press, New York, 1976.

12. **Seabrook, W. D.,** Neurobiological contributions to understanding insect pheromone systems, in *Annual Review of Entomology,* Vol. 23, Annual Reviews, Inc., Palo Alto, California, 1978, 471.

13. **Jacobsen, M.,** *Insect Sex Attractant Pheromones,* Academic Press, New York, 1972.

14. **Roelofs, W.,** Manipulating sex pheromones for insect suppression, in *Insecticides of the Future,* Jacobsen, M., Ed., Marcel Dekker, New York, 1975, 41.

15. **Brooks, T. W. and Kitterman, R. L.,** Gossyplure H. F. — pink bollworm suppression with male sex attractant pheromone released from hollow fibers — 1976 Experiments, *Proc. 1977 Beltwide Cotton Production-Mechanization Conf.,* National Cotton Council of America, Memphis, 1977, 79.

16. **Brooks, T. W. and Kitterman, R. L.,** Controlled Release Insect Pheromone Formulations Based on Hollow Fibers and Methods of Application, Paper No. 77-1035, presented at the Annual Meeting of the American Society of Agricultural Engineers, North Carolina State University, Raleigh, 1977.

17. **Metcalf, C. L., Flint, W. P., and Metcalf, R. L.,** *Destructive and Useful Insects Their Habits and Control,* McGraw-Hill, New York, 1962, 587.

18. **Van Steenwyck, R. A., Toscano, N. C., Ballmer, G. R., Kido, K., and Reynolds, H. T.,** Increases of *Heliothis* spp. in cotton under various insecticide treatment regimes, *Environ. Entomol.,* 4, 993, 1975.

19. **Finnel, C. M.,** private communication, 1977.

20. **Shorey, H. H., Kaae, R. S., and Gaston L. K.,** Sex pheromones of Lepidoptera, development of a method of pheromonal control of *Pectinophora gossypiella* in cotton, *J. Econ. Entomol.,* 67, 347, 1974.

21. **Green, N. M., Jacobsen, M., and Keller, J. C.,** Hexalure, an insect attractant discovered by imperical screening, *Experientia,* 25, 682, 1969.

22. **Keller, J. C., Sheets, L. W., Green, N., and Jacobsen, M.,** *Cis*-7-hexadecen-1-ol acetate (hexalure), a synthetic sex attractant for pink bollworm males, *J. Econ. Entomol.,* 62, 1520, 1969.

23. **Shorey, H. H., Gaston, L. K., and Kaae, R. S.,** Air permeation with gossyplure for control of the pink bollworm, in *Pest Management with Insect Sex Attractants,* Beroza, M., Ed., ACS Symposium Series 23, American Chemical Society, Washington, D. C., 1976, 67.

24. **Hummel, H. E., Gaston, L. K., Shorey, H. H., Kaae, R. S., Byrne, K. J., and Silverstein, R. M.,** Clarification of the chemical status of the pink bollworm sex pheromone, *Science,* 181, 873, 1973.

25. **Bierl, B. A., Beroza, M., Staten, R. T., Sonnet, P. E., and Adler, V. E.,** The pink bollworm sex attractant, *J. Econ. Entomol.,* 67, 211, 1974.

26. **Gaston, L. K., Kaae, R. S., Shorey, H. H., and Sellers, D.,** Controlling the pink bollworm by disrupting sex pheromone communication between adult moths, *Science,* 196, 904, 1977.

27. **Roelofs, W. L., Tette, J., Taschenberg, E. F., and Comeau, A.,** Sex pheromone of the grape berry moth: identification by classical and electroantennogram methods and field tests, *J. Insect Physiol.,* 17, 2234, 1971.

28. **Taschenberg, E. F. and Roelofs, W. L.,** Pheromone communication disruption of the grape berry moth with microencapsulated and hollow fiber systems, *Environ. Entomol.,* 5, 688, 1976.

29. **Taschenberg, E. F. and Roelofs, W. L.,** Mating disruption of the grape berry moth, *Paralobesia viteana* with pheromone released from hollow fibers, *Environ. Entomol.,* 5, 761, 1977.

30. **Roelofs, W. L., Comeau, A., Hill, A., and Milicevic, G.,** Sex attractant of the codling moth: characterization with electroantennogram technique, *Science,* 174, 297, 1971.

31. **McDonough, L. M. and Moffitt, H. R.,** Sex pheromone of the codling moth, *Science,* 183, 978, 1974.

32. **Carde, R. T., Baker, T. C., and Castrovillo, P. J.,** Disruption of sexual communication in *Laspeyresia pomonella* (codling moth), *Grapholitha molesta* (Oriental fruit moth), and *G. prunivora* (lesser appleworm) with hollow fiber attractant sources, *Entomol. Exp. Appl.,* 22, 280, 1977.

33. **Moffitt, H. R.,** unpublished studies, 1977.

34. **Weatherston, J., Roelofs, W., Comeau, A., and Sanders, C. J.,** Studies of physiologically active arthropod secretions. X. Sex pheromone of the eastern spruce budworm, *Choristoneura fumiferana* (Lepidoptera: Tortricidae, *Can. Entomol.,* 103, 1741, 1971.

35. **Sanders, C. J.,** unpublished studies, 1977

36. **Anon.,** The plight of the Tussock moth, *Environ. Sci. Technol.,* 8, 506, 1974.

37. **Smith, R. G., Daterman, G. E., and Daves, G. D.,** Douglas-fir tussock moth: sex pheromone identification and synthesis, *Science,* 188, 63, 1975.

38. **Sower, L. L. and Daterman, G. E.,** Evaluation of synthetic sex pheromone as a control agent for Douglas-fir tussock moths, *Environ. Entomol.,* 6, 889, 1977.

39. **Grant, G. G.,** unpublished studies, 1977.

40. **Daterman, G. E. and Sower, L. L.,** private communication, 1978.
41. **Cardarelli, N.,** *Controlled Release Pesticide Formulations,* CRC Press, Cleveland, 1976.
42. **Zweig, G.,** Environmental aspects of controlled release technology, in *Controlled Release Pesticides,* Scher, N. B., Ed., ACS Symposium Series 53, American Chemical Society, Washington, D. C. 1977, 37.
43. Federal Register, 42, 182, 47.205, Sept. 20, 1977.

Chapter 10

DELIVERY OF ACTIVE AGENTS BY OSMOSIS*

Felix Theeuwes

TABLE OF CONTENTS

*   Reprinted with permission from Goulding, Robert C., Ed., *Proc. Int. Controlled Release Pesticides Symp.*, Oregon State University, Corvallis, 1977, 364-381.

I. INTRODUCTION

Among the various forms of stored energy — such as electrical, elastic, electrochemical, mechanical, and chemical — that can be utilized to control delivery of active agents, the contributions based on osmosis are of particular interest.

Osmosis was discovered in 1748 by Nollet, and the first quantitative studies of osmotic pressure were published in 1887 by Peffer. The transport of liquids by osmosis was qualitatively discussed in 1896 by Starling, who identified osmotic and hydrostatic pressure differences across capillary membranes as important factors governing transcapillary fluid transport. Since these early discoveries, most of the progress has been in the field of reverse osmosis, where models describing transport processes as well as a large number of semipermeable membrane materials have been screened.

Chemical energy provides the driving force for osmotic flow and provides the means whereby solution diffusion transport of agents is accomplished across rate-controlling membranes. In the latter case, the agent is transported across the membrane under the influence of a thermodynamic activity gradient; however, in osmotic systems, water is transported by its thermodynamic activity gradient and is used to displace active agent through a properly designed delivery orifice.

Several benefits can be derived from osmotic mechanisms compared with solution diffusion transport of active agents across membranes.

1. Osmotic mechanisms provide the opportunity to produce generic systems that operate independent of agent properties.
2. Agents delivered can be of any molecular weight and chemical composition, including macromolecules and ionic species, which are usually difficult to deliver by solution diffusion.
3. Because of the small size of the water molecule compared to active agents, large delivery rates, compared to rates that can be obtained by agent solution diffusion, are expected. Using the solution diffusion mechanism, maximum normalized fluxes are found on the order of $0.2 \, \mu g / cm \cdot hr$.

II. DEFINITION OF VARIABLES AND FUNDAMENTAL EQUATIONS

Before discussing systems design, two fundamental working equations will be reviewed. The first equation relates escaping tendency of the solvent to osmotic pressure, which can then be considered as the driving force for solvent transport. The second equation describes solvent flow by osmosis. In Figure 1, a semipermeable membrane separates chambers 1 and 2 of the U-tube. Both chambers contain a common solvent (e.g., water) at different solute concentrations, such that the vapor pressure of water above solutions 1 and 2 are P_1^V and P_2^V, respectively. When both chambers contain identical solutes, the lower vapor pressure (P_2^V) is associated with the higher solute concentration. Water at a higher escaping tendency (P_1^V) from chamber 1 will diffuse across the semipermeable membrane to chamber 2 until the hydrostatic pressure difference, $P_{2,0}^H - P_{1,0}^H$, is reached, as given by Equation 1.

$$P_{2,0}^H - P_{1,0}^H = \pi_2 - \pi_1 = -\frac{RT}{V} \ln \frac{P_2^V}{P_1^V} \tag{1}$$

where R is the gas constant, T the absolute temperature, and V the molar volume of the solvent.

The hydrostatic pressure difference at equilibrium is defined as the osmotic pressure

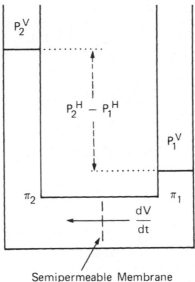

FIGURE 1. Osmosis cell.

difference, $\pi_2 - \pi_1$, which relates to the vapor pressure ratio, as shown in Equation 1.

Equation 2 describes volume flow, dV/dt, of water across the membrane when going toward equilibrium.[1]

$$\frac{dV}{dt} = \frac{A}{h} L_p \left[\sigma (\pi_2 - \pi_1) - \left(P_2^H - P_1^H \right) \right] \qquad (2)$$

where A is the membrane area, h the membrane thickness, L_p the mechanical permeability coefficient, $(P_2^H - P_1^H)$ the hydrostatic pressure difference across the membrane, and σ the reflection coefficient. The reflection coefficient characterizes the leakiness of the membrane — zero for a porous membrane and unity for an ideal semipermeable membrane.

Mass transport with osmotic systems is obtained by displacement of the agent at a concentration, C, obtained through systems design at the volume flow described by Equation 2. The magnitude of the delivery rate can be designed through engineering of the parameters defined in Equation 2 and the driving force defined by Equation 1, which can further be estimated from Van t'Hoff's law:

$$\pi = C \cdot R \cdot T \qquad (3)$$

where C is the concentration of osmotic solute, and π, R, and T as previously defined.

III. EARLY SYSTEMS DEVELOPMENT

In 1955, Rose and Nelson[2] developed a continuous long-term injector based on osmotic principles and claimed to have made devices that delivered 0.02 ml/day for 100 days, up to 0.5 ml/day for 4 days. The system (Figure 2a) was developed as a tool for pharmacologic research.[3] The drug is contained in the space between the glass ampule (D) and a latex bag (B). Congo red as the osmotic driving agent and a stainless steel ball as the stirring system are contained within the bag, which is closed off by a

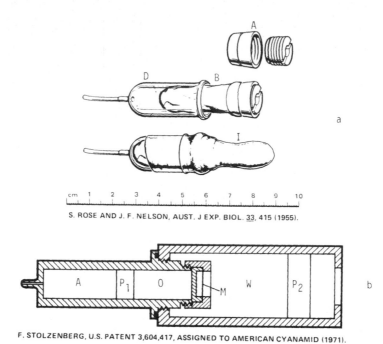

S. ROSE AND J. F. NELSON, AUST. J EXP. BIOL. 33, 415 (1955).

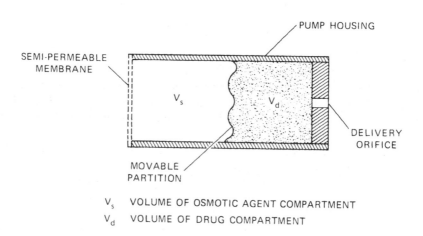

F. STOLZENBERG, U.S. PATENT 3,604,417, ASSIGNED TO AMERICAN CYANAMID (1971).

FIGURE 2.   Early osmotic system development. (From Rose, S. and Nelson, J. F., *Aust. J. Exp. Biol. Med. Sci.*, 33, 415, 1955. With permission.)

V_s VOLUME OF OSMOTIC AGENT COMPARTMENT
V_d VOLUME OF DRUG COMPARTMENT

FIGURE 3.   Schematic representation of an osmotic pump.

cellophane membrane clamped in a holder (A). The system contains its own water supply in a second bag (I).

The first patent on an osmotic pump was filed by Stolzenberg[4] in 1971 (Figure 2b). The system contains the agent in a chamber (A) separated by a piston (P_1) from the osmotic solution (O). The semipermeable membrane (M) separates the osmotic solution from the water in the chamber (W), which is contained by a piston (P_2) and separated from the environment.

Both systems are of the type schematically shown in Figure 3. The movable partition represents either the latex bag (Figure 2a) or the piston (Figure 2b). The delivery rate from such a system is given by Equation 4.

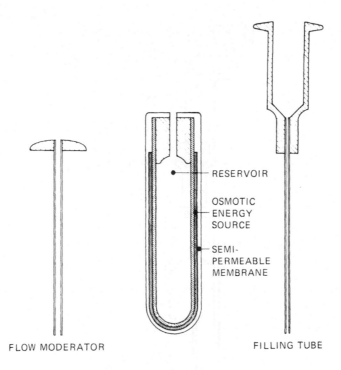

FIGURE 4.   Osmotic pump and components.

$$\frac{dm}{dt} = \frac{dv}{dt} \cdot C_S \tag{4}$$

where C_s is the concentration of the active agent in the chamber, V_d. The volume flow (dV/dt) produced by the osmotic imbibition of water into the chamber containing the osmotic driving agent is given by Equation 2. The hydrostatic pressure difference ($P_2^H - P_1^H$) across the semipermeable membrane in the system represented by Figure 3 is generated by the stiffness of the partition and the rheology of the formulation in the compartment, V_d, in conjunction with the delivery orifice.

IV. MINI-OSMOTIC PUMP

A. Description

The mini-osmotic pump[5] is a system of the type shown by Figure 3, constructed according to a layering technique, where optimum usage of space and minimum quantities of materials are employed. The system is shown in cross section (Figure 4) together with the filling tube and the flow moderator, which is inserted into the system after filling. The system is fabricated and sold* unfilled and programmed at a volume delivery rate fixed by selection of the osmotic driving agent and the membrane. The pump consists of an inner collapsible reservoir surrounded by the osmotic driving agent, potassium sulfate. The driving agent is, in turn, surrounded by a cellulosic semipermeable membrane that makes a leakproof seal with the elastomeric reservoir at the top. The total system is approximately 2.5 cm long and has a diameter of 0.65 cm. The reservoir volume is 170 $\mu\ell$.

---

*   ALZET® mini-osmotic pump, ALZA Corporation, Palo Alto.

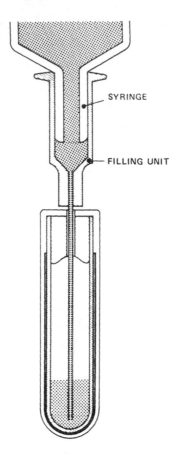

FIGURE 5.   Filling the osmotic pump.

B. Delivery Rate

The volume flow, as derived from Equation 2 for these systems,[5] is described by Equation 5.

$$\frac{dV}{dt} = k\frac{A}{h}\ (\pi_S - \pi_e) \tag{5}$$

where $k = L_p \sigma$. For the cellulosic membrane selected, $\sigma = 1$, using potassium sulfate as the driving agent. $\pi_S = 40$ at is the osmotic pressure of the potassium sulfate that is constant by providing an excess of salt throughout the operational lifetime of the pump. The osmotic pressure of the surrounding environment in which the pump will be operating is π_e. The hydrostatic pressure difference is zero through the design of a sufficiently large orifice and the selection of the flexible bag. Although the volume delivery rate is programmed, the mass delivery rate, Equation 4, can be adjusted by filling the unit at any concentration, C_s. The system is filled as shown in Figure 5 and, after installation of the flow moderator, is shown in operation in Figure 6. The flow moderator is installed mainly to reduce the amount of agent delivered from the system by diffusion through the orifice.

As an example, the cumulative volume delivered from the system is shown in Figure 7; the system pumps 1 μl/hr in isotonic saline or in the s.c. space of test animals where water from the environment displaces the agent.

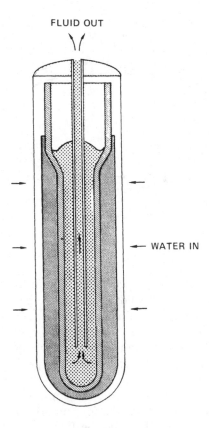

FIGURE 6.   Osmotic pump in use.

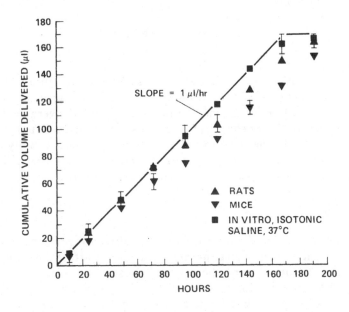

FIGURE 7.   Cumulative volume delivered from miniaturized osmotic pump.

C. Applications

As seen from Figure 7, good correlation is found between the in vitro rate in isotonic

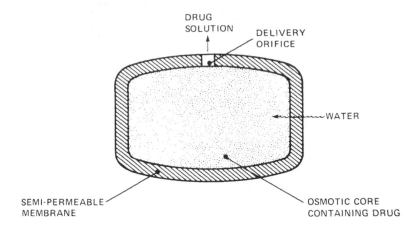

FIGURE 8.   Elementary osmotic pump cross section.

saline at 37°C and delivery rates obtained as a s.c. implant in mice and rats. Thus, the system is particularly useful in the study of pharmacology of drug substances.

Furthermore, the system can be used as a screening tool in any application where water is available at a higher activity than the activity in equilibrium with the saturated salt solution used as the driving agent in the pump. Such applications include delivery of active agents (e.g., growth promoters, algicides, pesticides, and fungicides) to streams, ponds, soil, and plants.

In addition, the system can be used in dedicated applications where it is filled before application or manufactured with the desired formulation. For dedicated applications, a simpler system is available, one which can be fabricated with an agent composing up to 95% of the total volume. This system, also based on osmotic pump technology, is called the elementary osmotic pump.

V. ELEMENTARY OSMOTIC PUMP

A. Description

The elementary osmotic pump[6] is shown in cross section in Figure 8. It consists of an active agent, optionally admixed with an osmotic driving agent, as the solid core and surrounded by a semipermeable membrane containing a delivery orifice. The system can be made to any size or shape specifications. The delivery rate is programmed by the osmotic pressure of the core formulation and by the permeability of the semipermeable membrane to water. The delivery rate is independent of the orifice size when the orifice cross-sectional area, A_O, is contained within two critical limits.[7]

$$A_{min} \leq A_O \leq A_{max} \tag{6}$$

The area, A_o, must be sufficiently large to prevent hydrostatic pressure build-up within the system, which can deform the membrane enclosure or affect the pumping rate, as shown by Equation 2. Furthermore, the area, A_o, must be sufficiently small to prevent delivery of the agent by diffusion through the fluid medium contained in the orifice. For systems designed with a delivery rate on the order of 10 mg/hr, the diameter of a single cylindrical orifice is typically about 0.2 mm through a membrane 0.2 mm thick.

B. Delivery Rate

Delivery of the agent from these systems is accomplished by water imbibition from the environment, at a rate described by Equation 2. The water instilled into the system displaces an equal volume of agent formulation created *in situ* such that the delivery rate of the system can be described by Equation 7.

$$\frac{dm}{dt} = k \cdot \frac{A}{h} \left(\pi_{eff} - \pi_e \right) \cdot C \tag{7}$$

The hydrostatic pressure difference shown in Equation 2 can be neglected compared to the osmotic driving pressure, $(\pi_{eff} - \pi_e)$, as obtained through proper a orifice design discussed above. The parameters, k, A, h, and π_e, are as previously defined, and π_{eff} is the effective osmotic pressure created by the agent formulation. The concentration of agent, C, of the formulation dispensed can be either in solution or suspension.

When the agent is also the osmotic driving agent and the sole compound contained within the system, the delivery rate is constant as long as excess solid is present within the membrane compartment. The elementary osmotic pump, under this condition, will pump at a constant rate since its volume remains essentially constant, and a constant influx of water is translated into a constant outflow of saturated solution. When the osmotic pressure of the saturated solution dispensed is large compared with the osmotic pressure of the environment, the zero order rate is given by Equation 8.

$$\left(\frac{dm}{dt} \right)_z = k \cdot \frac{A}{h} \cdot \pi_s \cdot S \tag{8}$$

where S is the agent solubility and π_s is the osmotic pressure of its solution. When all of the solid agent has been taken up in solution at time, t_z, the delivery rate will decline parabolically because both the osmotic pressure and the concentration of the solution decline as a function of time. It has been shown that the nonzero order rate can be described by Equation 9.

$$\frac{dm}{dt} = \frac{(dm/dt)_z}{\left[1 + \frac{1}{S \cdot V} \left(\frac{dm}{dt} \right)_z (t - t_z) \right]^2} \tag{9}$$

where V is the internal volume of the membrane compartment. As an example, the experimental and theoretical rates, calculated from Equations 8 and 9, are compared in Figure 9. Close agreement, of the type shown in Figure 9, between experiment and theory has been found for a large number of systems, indicating the validity of the osmotic mechanism presented.

Important properties of the delivery rate profile that are expected on theoretical grounds and have been verified experimentally are

1. The delivery rate of the system is high when normalized to surface area and thickness. For example, the normalized zero order rate for the systems (Figure 9) with a surface area of 2.2 cm^2 and a membrane thickness of 0.025 cm is 250 μg/cm·hr compared to a transference[7] on the order of 0.2 μg/cm·hr obtained by solution diffusion.
2. Since highly water-soluble agents can be delivered and the agent solution is pumped away from the membrane surface area, avoiding concentration polarization, the delivery rate from the system tends to be independent of the stirring rate (Figure 10).

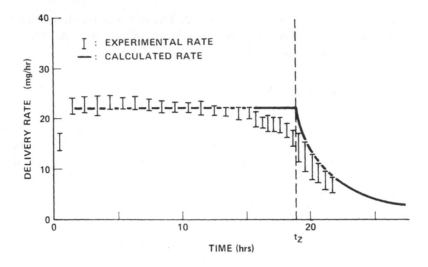

FIGURE 9.   In vitro release rate of potassium chloride from elementary osmotic pumps in water at 37°C. Key: I range of experimental data obtained from five systems; —, calculated release rate.

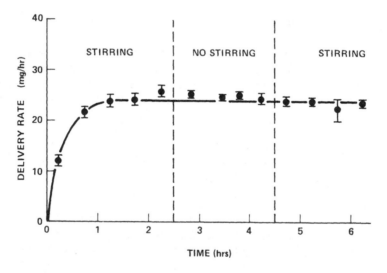

FIGURE 10.   In vitro release rate of potassium chloride from elementary osmotic pumps in water at 37°C. The vertical dashed lines indicate the time at which the systems were transferred from a stirred to a stagnant medium and back to a stirred medium. I represents the range of experimental data obtained from five systems.

3.   The formulation inside of the elementary osmotic pump system is shielded from the environment and can be formulated at a predetermined pH. Ion exchange across the membrane does not take place since the membrane is semipermeable and only water is instilled into the system. The delivery rate from the system is, therefore, independent of pH, although the solubility of the agent may be strongly dependent on pH. One example is a system delivering phenobarbital (Figure 11) at pH = 1.2 (artificial gastric fluid) for 2 hr and at pH = 7.5 thereafter.

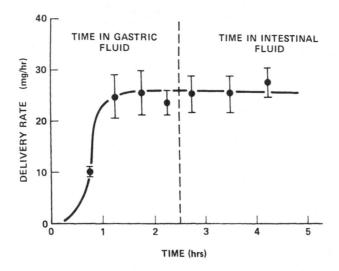

FIGURE 11. In vitro release rate of sodium phenobarbital from elementary osmotic pump systems in gastric and intestinal fluid USP without enzymes. The vertical dashed line indicates the time at which the systems were transferred from gastric to intestinal fluid. I represents the range of experimental data obtained from three systems.

C. Applications

The elementary osmotic pump is well suited for dedicated applications where the system can be preprogrammed to deliver the agent at a specified rate and water is available at sufficiently high activity. Fields of application include

1. Pharmaceutical and medical: for the delivery of agents (or combinations of agents) in any body cavity
2. Agricultural: for the delivery of fertilizers, growth promoters, or pesticides
3. Horticultural: for the local or systemic delivery of nutrients or pesticides to plants
4. Water treatment: for the delivery of agents to ponds and streams

REFERENCES

1. **Katchalsky, A. and Curran, P. F.**, *Non Equilibrium Thermodynamics in Biophysics,* Harvard University Press, Cambridge, Mass., 1976.
2. **Rose, S. and Nelson, J. F.**, A continuous long-term injector, *Aust. J. Exp. Biol. Med. Sci.,* 33, 415, 1955.
3. **Urguhart, J. Davis, J. O., and Higgins, J. T., Jr.**, Simulation of spontaneous secondary hyperaldosteronism by intravenous infusion of agiotension II in dogs with an arterio venous fistula, *J. Clin. Invest.,* 43, 1355, 1964.
4. **Stolzenberg, S. J.**, Osmotic Fluid Reservoir for Osmotically Activated Long-term Continuous Injector Device. U.S. Patent 3,604,417. Patented: Sept. 14, 1971. Assignees: American Cynamid Company, Stamford, Conn.
5. **Theeuwes, F. and Yum, S. I.**, Principles of the design and operation of generic osmotic pumps for the delivery of semisolid or liquid drug formulations, *Ann. Biomed. Eng.,* 4, 343, 1976.
6. **Theeuwes, F.**, Elementary osmotic pump, *J. Pharm. Sci.,* 64, 1987, 1975.
7. **Theeuwes, F., Gale, R. M., and Baker, R. W.**, Transference: a comprehensive parameter governing permeation of solvents through membranes, *J. Membr. Sci.,* 1, 3, 1976.

Chapter 11

STARCH AND OTHER POLYOLS AS ENCAPSULATING MATRICES FOR PESTICIDES

B. S. Shasha

TABLE OF CONTENTS

I. GENERAL DESCRIPTION

Several forms of controlled release pesticides have been reported, ranging from formulation in Attaclay granules,[1] plaster of paris,[2] and polymeric granules[3] to systems where the pesticide is covalently bonded to lignocellulosic wastes[4] and starch.[5] A simple and economical way to entrap water-insoluble compounds, including pesticides, within a starch matrix is described here.[6] The procedure consists of dispersing the active agent in an aqueous starch xanthate solution and, subsequently, crosslinking the starch xanthate either oxidatively, with multivalent metal ions, or with difunctional reagents such as epichlorohydrin. Cereal flours, which contain about 10% protein along with starch, also can be xanthated and used as an encapsulating matrix. Upon crosslinking, which is effected within a few seconds under ambient conditions, the entire mass becomes gell-like and on continued mixing for an additional few seconds becomes a particulate solid that can be dried to a low moisture content with only minimal or no loss of the entrapped chemical.

Starch[7] is one of the most abundant biopolymers. It is found mainly in the plant kingdom where it occurs as the principal food reserve, polysaccharide. Currently, over 200 billion lb are produced in this country by the corn crop alone. Besides availability, it is a very inexpensive polymer; current prices of corn starch in bulk quantities run less than 10 cents per pound. It consists of glucopyranose units of α-D-(1 $\rightarrow$ 4) linkages (Figure 1) and hydrolyzes completely to yield D-glucose. It can be easily fractionated by chemical means to yield amylose, a straight-chain polymer, and amylopectin, a branched-chain polymer. While amylose is capable of forming tough flexible films, amylopectin does not possess this property. Starch reacts with iodine solution to give a dark-blue color.

Because of the abundance of free hydroxyls, starch is easily derivatized. The degree of derivatization or substitution (D.S.) is defined as the number of substituents per glucose unit. Thus, theoretically, a D.S. of three is possible. For the purpose of encapsulation, starch is derivatized through xanthation to yield a D.S. of 0.1 to 0.3, followed by crosslinking to yield starch xanthide. The chemistry and technology of xanthates,

$$\text{Starch} - \text{OH} + \text{CS}_2 + \text{NaOH} \longrightarrow \text{Starch} - \overset{\overset{\text{S}}{\|}}{\text{OCSNa}}$$

$$\xrightarrow[\text{}]{\text{H}^+, [\text{O}]} \text{Starch} - \overset{\overset{\text{S}}{\|}}{\text{OCS}}\overset{\overset{\text{S}}{\|}}{\text{SCO}} - \text{Starch}$$

especially cellulose xanthate, is well documented.[8] In recent years, crosslinked starch xanthate has been shown to have potential as a paper additive,[9] for reinforcement of rubber,[10] for making powdered rubber,[11] and for removing heavy metals from aqueous solutions.[12]

II. FACTORS AFFECTING RATE OF RELEASE*

Shelf life of the starch-encapsulated pesticides is good, and there is no appreciable loss on storage during at least 1 year. When placed in open containers for several weeks, loss of the volatile agent is negligible. However, when products are wetted or immersed in water, the active agent is then released from the matrix. A simple labora-

---

* Mention of a pesticide in this paper does not constitute a recommendation for use by the U.S. Department of Agriculture nor does it imply registration under FIFRA, as amended.

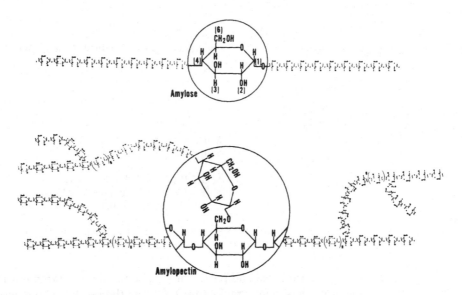

FIGURE 1.   Structural features of starch components.

tory screening test (wet test) was devised for comparing the release properties of thio-carbamate-containing products to assist in the selection of formulations for subsequent bioassay. Other factors that affect the rate of release are the characteristics of the entrapped chemical, e.g., in an aqueous medium, the higher the solubility in water of the active agent, the faster it will diffuse out of the starch matrix. Thus, the herbicide S-ethyl dipropylthiocarbamate (EPTC) will be released faster than the herbicide S-ethyl diisobutylthiocarbamate (butylate). (Water solubility of EPTC at 20°C is 370 ppm and of butylate at 22°C is 45 ppm.)

The use of polymeric materials in controlled release application systems almost always involves consideration of the solubility and diffusivity of the active agent in the polymer matrix.[13] The literature deals extensively with the solution, diffusion, and permeation of low-molecular-weight gases, vapors, liquids, and ions in polymer films.[14-19]

The characteristics of the polymeric xanthate matrix used for the encapsulation also play an important role in the rate of release. It was found,[20] for example, that, under moist conditions, an acid-modified flour matrix releases active agent faster than an acid-modified flour-starch mixture and that this mixture releases faster than starch alone.

It has been suggested[20] that either the protein component in the flour or the lower molecular weight of the starch component of the flour contributes to a faster release of butylate. Also, the recovery of the encapsulated agent is higher with starch than with the acid-modified flour-starch mixture and poorest with acid-modified flour alone (Table 1).

The addition of small amounts of latex, such as SBR 1502, to the xanthate retarded significantly the rate of release of 1,2-dibromo-3-chloropropane (DBCP) (Table 2). The addition of predissolved polymers, such as polystyrene in benzene, seems to have the same effect. On the other hand, the rate of release of butylate or of EPTC was not changed significantly by the addition of latex SBR 1502 nor by the addition of polystyrene.

When made with xanthates having a D.S. of 0.3 and with H_2O_2 as an oxidant, the products release the active agent more slowly than do those made with the same D.S., but with $NaNO_2$ as an oxidant. The reason seems to be that with $NaNO_2$, during the

TABLE 1

Recovery Rate of Butylate[a] Formulations

Xanthate base[b]	Recovery rate (%)
Acid-modified starch	<20
Acid-modified flour	<20
Acid-modified starch: starch (1:1 ratio)	61
Acid-modified flour: starch (1:1 ratio)	60
Starch	80

[a]   S-ethyl diisobutylthiocarbamate.
[b]   Xanthate D.S. was 0.3 and contained about 13% solid; H_2O_2 was used as oxidant.

TABLE 2

Release Properties of DBCP[a] Formulations[b]

Formulation	Loss of DBCP (%)	
	20 hr	72 hr
Starch xanthide[c]-latex[d]	15	26
Starch xanthide	30	54

[a]   1,2-Dibromo-3-chloropropane.
[b]   Rate of release was monitored by wet test.
[c]   Xanthate D.S. was 0.35; H_2O_2 was used for oxidation.
[d]   Latex SBR 1502 was added at 20% level, based on starch (solid/solid basis).

neutralization step, NO_2 gas is produced and partly entrapped within the starch matrix. Upon drying, the end product has many cracks that facilitate the release of the active ingredient. With H_2O_2, on the other hand, the product has a smooth and continuous surface (Figure 2).

The technique of double encapsulation provides a slower rate of release with certain pesticides than does single encapsulation, especially in cases where a single encapsulation does not entrap all of the active agent and some of it is adsorbed loosely to the surface of the starch matrix. The double encapsulation technique involves the addition of another layer of xanthate to the crosslinked xanthate containing the active agent. After mixing, the second layer is crosslinked as before.

Finally, the mesh sizes of the end product also seem to influence the rate of release. Preliminary data show that EPTC and DBCP with granular sizes of mesh 30 or higher have a rate of release faster than a product with granular size of mesh lower than 30 (Table 3 and Table 4).

III. METHODS FOR TESTING ENCAPSULATED PRODUCTS

1.   If the encapsulated agent is a solid with low solubility in common solvents, then the product is examined through a magnifying lens with 30 to 50 times magnification to observe whether loose particles of the agent are present unencapsulated.

 A second test which is more meaningful and easy to carry out involves immersion of the product in a diluted aqueous iodine solution for 1 to 2 min. Since only the starch part will stain with iodine, the observation of the immersed sample through the lens will reveal the quality of entrapment.

2.   If the encapsulated agent is a solid or a liquid soluble in common organic solvents, then, besides the iodine test, a portion of the product (about 10 g, containing 15% encapsulated agent) is suspended in the appropriate organic solvent (50 mℓ) for 1 hr. The portion of the agent extracted reflects the part that is not encapsulated or that is close to the surface of the encapsulated granule.

3.   If the encapsulated agent is a volatile liquid, a portion of the product is exposed to an air current (well-ventilated hood will suffice), and the remaining product is analyzed periodically. Another simple test, the wet test, consists of placing several 1-g portions of the product in 2 mℓ of water. The water is allowed to evaporate, and, after 24 hr, another 2 mℓ of water is added and again allowed to evaporate. The addition of water and evaporation is repeated a total of four times, and the

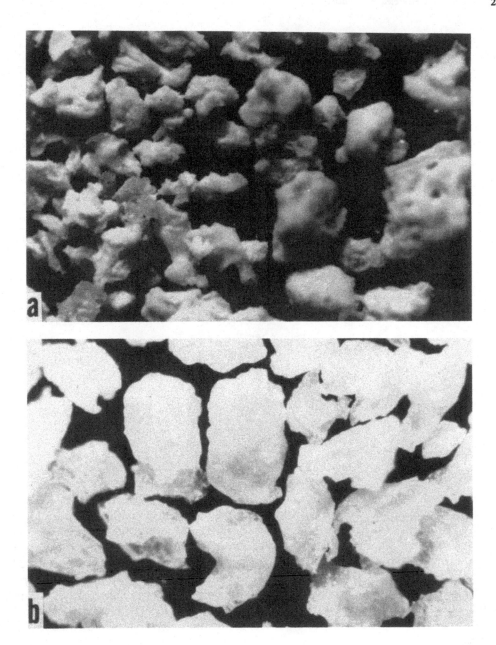

FIGURE 2.   Starch xanthide granules prepared with different oxidants. (a) NaNO₂, (b) H₂O₂.

dried product is analyzed after each wetting-drying cycle for loss of encapsulated agent.

IV. LARGE BATCH PREPARATIONS

A number of pesticides have been encapsulated in starch xanthate matrix by a batch-type process in 2- to 11-lb quantities. Recovery of the pesticides as the encapsulated products was 68 to 100%. The products contained approximately 15 to 25% active pesticide.

The samples listed in Table 5 were prepared by crosslinking starch xanthate of D.S.

TABLE 3

Properties of Starch Xanthide-EPTC[a] Formulations as Related to Particle Size[b]

Mesh size	EPTC (%)	Loss of EPTC after 2 days water treatment (% of total)
>60	14.0	14
30—60	21.7	17
14—30	21.7	5
<14	21.7	5

[a] S-ethyl dipropylthiocarbamate.
[b] Starch xanthate D.S. was 0.175 and crosslinked with H_2O_2. The product after drying was separated into fractions by sieving.

TABLE 4

Properties of Starch Xanthide-DBCP Formulations as Related to Particle Size[a]

Mesh size	Loss of DBCP after 4 days aeration (% total)
60	7.0
40	4.0
20	0

[a] Acid-modified flour xanthate D.S. was 0.3 and crosslinked with H_2O_2. The product after drying was separated into fractions by sieving.

0.3 with hydrogen peroxide, using the double encapsulation procedure. The recovery of the encapsulated pesticides ranged from 80 to 95%. Additional pesticide could be recovered from the filtrate by allowing the filtrate to separate into two phases and decanting or by extracting the filtrate with an appropriate solvent, such as hexane or carbon tetrachloride. The products all exhibited slow-release properties, as indicated by the wet test (Table 5). Results of wet tests on earlier laboratory preparations correlated well with results from greenhouse evaluations. Laboratory preparations of encapsulated pesticides which lost less than 20% of the active agent in 96 hr were considered to have slow-release properties. The wet test results in Table 5 show a 5 to 20% loss for these products.

The pesticides were also singly encapsulated with starch xanthate of D.S. 0.3 with 76 to 83% recovery of active agents, which is slightly lower than for the doubly encapsulated products (Table 6). Wet test results indicated release properties should be about the same as for the doubly encapsulated products.

Starch xanthates of D.S. 0.2 were also used for encapsulation. The recovery, shown in Table 7, ranged from 74 to nearly 100%. Wet test results of these products were generally lower than those of the higher D.S. products, which suggests that these products should have slower release properties.

Table 8 shows the loss of pesticide from a physical mixture of starch xanthide and each of four pesticides, added at the 20% level and treated with water as before.

The products in Tables 5 to 8 were prepared by crosslinking the xanthate with hydrogen peroxide. In addition, products have been prepared using other crosslinking

TABLE 5

Double Encapsulation of Pesticide Emulsifiable Concentrates Using Starch Xanthate of a D.S. of 0.3

Pesticide	% Active in product	% Active recovered	Wet test (% loss, 96 hr)
Butylate	18.9	86	5
Butylate-plus	14.9	95	—
EPTC	21.3	88	19
Vernolate	18.0	82	—

TABLE 6

Single Encapsulation of Pesticide Emulsifiable Concentrates Using Starch Xanthate of a D.S. of 0.3

Pesticide	% Active in product	% Active recovered	Wet test (% loss, 96 hr)
Vernolate	22	76	1
Butylate	23	81	7
EPTC	23.8	83	14

TABLE 7

Single Encapsulation of Pesticide Emulsifiable Concentrates Using Starch Xanthate of a D.S. of 0.2

Pesticide	% Active in product	% Active recovered	Wet test (% loss, 96 hr)
Cycloate	22.0	84	0
Butylate	21.4	90	3
Butylate-plus	19.5	85	5
EPTC	24.0	89	12
Vernolate	22.5	96	4
Surpass	20.0	74	9
Pebulate	18.0	90	3
Dimilin®	10	100	—
EPTC-plus	16	81	13

TABLE 8

Loss of Pesticide from Physical Mixtures with Starch Xanthide

Pesticide	Vapor pressure[a]	Pesticide lost (%) 24 hr	72 hr
EPTC	34×10^{-3}	85	94
Cycloate	6.2×10^{-3}	15	32
Vernolate	10.4×10^{-3}	40	89
Butylate	13×10^{-3}	75	96

[a] At ambient temperature.

TABLE 9

Encapsulation of Pesticide Emulsifiable Concentrates with Starch Xanthate and Different Oxidants

Pesticide	Xanthate D.S.	Crosslinking agent	% Active in product	% Active recovered
EPTC-plus	0.20	NaOCl	15	62
EPTC	0.12	FeCl$_3$	27	97
EPTC	0.12	NaNO$_2$	22	66
EPTC	0.12	NaClO$_2$	25	74
Picloram	0.30	H$_2$O$_2$	9	98

agents (Table 9). Xanthates of D.S. 0.3, when crosslinked with H$_2$O$_2$, gave firm hard granules after grinding in the Bauer mill and were easily dried. Xanthates of D.S. 0.2 gave softer particles which contained more water and tended to stick together as the samples dried making them more difficult to handle and slower to dry. Xanthates of D.S. less than 0.15 retained too much water to be usable when H$_2$O$_2$ was the oxidant. However, if NaClO$_2$, FeCl$_3$, or NaNO$_2$ were used for crosslinking, D.S. of 0.15 products had physical properties similar to the D.S. 0.3, H$_2$O$_2$ crosslinked product.

Encapsulation of EPTC-plus was made in which the filtrate from the previous run was used for the makeup water in the xanthation step. No difficulties were encountered. Xanthation proceeded normally, although the color was a much brighter yellow. The encapsulated product retained 76% of the pesticide. Another 15% of the active agent was extracted from the filtrate, making a total recovery of 91%.

As for equipment, a sigma-blade stainless steel mixer reactor of 10-ℓ working capacity was used for xanthation and encapsulation reactions and yielded 1 ½ to 2 ½ lb of product. Alternatively, a stainless steel jacketed 1 ½-ft^3 ribbon blender, equipped with a double ribbon agitator and a pneumatically controlled discharge gate, was used for preparing 6.5 to 11 lb. of product. An 8-in. Bauer mill, model no. 148-2, equipped with a no. 8319 plate, was used to grind the still-wet crosslinked starch xanthide encapsulated products (50 to 75% H$_2$O) to afford particle sizes in the 14 to 30 mesh range. All drying was done at ambient temperature.

V. RESULTS OF TESTS

The data obtained through the use of encapsulation within a starch matrix are encouraging and indicate considerable promise for this new technology. It should be understood that most of the data are preliminary and that minor changes in the formulation tend to change considerably the rate of release and, hence, the results of these tests.

Feldmesser et al.[21] reported on laboratory evaluations of starch xanthide formulations of DBCP, a nematicide, and Diazinon® [O,O-diethyl O-(2-isopropyl-6-methyl-4-pyrimidinyl)] phosphorothioate, an insecticide-nematicide for nematicidal activity. Two DBCP formulations containing 35.6 and 42.0% active agent and two Diazinon® formulations containing 42.0 and 43.6% active were studied. To determine retention of the active agent under various conditions, the products were aerated, both wet and dry, in open dishes for several days, and then the amount retained was determined by bioassays against nematodes in a standard in vitro test. During a 10-day aeration of wet DBCP formulations, most of the active agent was lost, and their subsequent effectiveness for killing nematodes was minimal. Retention of Diazinon® was considerably greater, perhaps due to its much lower vapor pressure than DBCP. During the 10-day

TABLE 10

Release Properties of Starch Xanthide-Diazinon® Formulation

	Percent kill			
	42.0% active agent		43.6% active agent	
	10 days	44 days	10 days	44 days
Starch xanthide-diazinon				
Wet[a]	69.4	63.0	76.3	52.5
Dry	66.2	63.1	75.4	70.0
Unaerated[b]	74.6	76.0	78.1	74.3
Technical diazinon				
Wet[a]	81.0	0	82.8	0
Dry	76.6	0	77.4	0
Unaerated[b]	89.0	0	95.1	0

[a] Second aeration period of 34 days, following 10-day wet and dry aeration and 48-hr exposure, was dry for all granules.

[b] Granules not aerated for first 10 days; Aerated dry for 34 days.

aeration tests and an additional 34 days of aeration in the dry, the two diazinon formulations lost only about 20 to 33% of their active ingredient. The mortality of *Panagrellus rediviuus*, a saprophagous nematode, after a 48-hr exposure to 200 ppm Diazinon® in the form of the two starch xanthide formulations, after various periods of aeration, is shown in Table 10.

It was also shown[21] that the nematicide loss through the starch matrix is mediated by moisture and that moist granules may be dried to regain their retentive capacity. Thus, such granules may retain and release nematicide during several wetting and drying cycles. These properties make it possible to apply such materials to soil with minimal loss of the active fractions to the atmosphere.

In a second report, Feldmesser and Shasha[22] studied the effects of DBCP on *Meloidogyne incognita*. The nematicide was applied to soil surfaces as technical material, EC (emulsifiable concentrate), and encapsulated in the starch matrix. The application was done 4 days before nematode inoculation and planting of tomato seedlings. The results (Table 11) show that the technical DBCP and EC DBCP treatments were ineffective at each of the two dose rates after 7 weeks, and no infections were comparable to those in the inoculated untreated controls. The encapsulated DBCP reduced root-knot infections by an average 70% at the lower dosage and by 95 to 100% for the higher dosage. None of the tomato plants showed symptoms of chemotoxicity.

In another report, four starch-encapsulated formulations of EPTC and of butylate were prepared and evaluated for their slow-release capabilities and efficacies compared to their respective EC formulations. Chemical (Tables 12 and 13) and biological (Tables 14 and 15) evaluation indicate that differences in controlled release can be achieved by selection of the starch xanthate and oxidant used in the formulation process. EPTC and butylate, formulated as starch-encapsulated granules, release slower than their respective ECs under soil conditions that favor rapid release. The initial release was adequate for weed control and slow enough for a desired residual.

Whereas many herbicides must be incorporated in the soil soon after application to the surface to prevent extensive losses due to volatilization or decomposition by sunlight, it is hoped that controlled release formulations may prolong the time before incorporation is needed or, ideally, to provide control without being incorporated. Four formulations containing butylate were evaluated for the ability to delay time

TABLE 11

Nematicidal Effects of DBCP Applied as 20% Active
Starch Xanthide Granules and as Tech (technical) and
EC

Treatment	Test 1	Test 2
Uninoculated control	0.0[a]	0.0
Inoculated control	4.0	3.7
Tech-DBCP		
9.5 kg ai/ha[b]	4.0	4.0
17.0 kg ai/ha	4.0	4.0
EC-DBCP		
9.5 kg ai/ha	4.0	4.0
17.0 kg ai/ha	4.0	4.0
Starch xanthide-DBCP granules		
9.5 kg ai/ha	1.5	1.0
17.0 kg ai/ha	0.5	0.0

[a] Root-knot index: 0 = nematode-free, 1.0 = 25% of
roots infected, 2.0 = 50%, 3.0 = 75%, and 4.0 =
100%.
[b] Active ingredient per hectare.

TABLE 12

Effect of Encapsulation Conditions on Rate of Release of Double Encapsulated EPTC as Determined
by Chemical Analysis

Formulation code	Type of formulation	Oxidant	Active ingredient (%)	Dry test; loss after 17 days (%)	Wet test; loss after (%)	
					48 hr	96 hr
60	Acid-modified flour + starch xanthate, 22.7% solid, D.S. of 0.3	NaNO₂	11	40	Values meaningless; compound caked under aqueous conditions	
61	Acid-modified flour + starch xanthate, 22.7% solid, D.S. of 0.3	H₂O₂	14.5	10	12	30
59	Starch xanthate, 11.5% solid, D.S. of 0.3 in a mixture with polystyrene	H₂O₂	16.6	3	18	40
58	Starch xanthate, 11.5% solid, D.S. of 0.3	H₂O₂	17.5	0	0	0

Note: As a reference point, starch xanthide mixed with an EC of EPTC at a rate of 15% lost over
95% of the ai in 48 hr in both the dry and the wet test.

before incorporation. Butylate, as an EC, was applied as a control. Rates of active
agent (ai) of 3 and 4 lb/acre were applied. Formulations were placed on the surface
of wet soil and incorporated immediately and after 24 hr. Results are shown in Tables
16 and 17.

Schreiber[25] reported results of greenhouse and field tests of two starch xanthide-
EPTC products. The formulations were made from starch xanthates of a D.S. of 0.35,
and sodium nitrite was used for crosslinking. One formulation contained 14% EPTC

TABLE 13

Effect of Encapsulation Conditions on Rate of Release of Double Encapsulated Butylate Determined by Chemical Analysis

Formulation code	Type of formulation	Oxidant	Active ingredient (%)	Dry test; loss after 17days (%)	Wet test; loss after (%)		
					24 hr	48 hr	96 hr
68	Acid-modified flour + starch xanthate, 22.7% solid, D.S. of 0.3	NaNO$_2$	14.5	60	29	68	68
69	Acid-modified flour + starch xanthate, 22.7% solid, D.S. of 0.3	H$_2$O$_2$	15.2	29	20	36	52
67	Starch xanthate, 11.5% solid, D.S. of 0.3 in a mixture with latex	H$_2$O$_2$	18.6	15	0	0	22
66	Starch xanthate, 11.5% solid, D.S. of 0.3	H$_2$O$_2$	18.7	0	0	0	0

Note: As a reference point, starch xanthide mixed with an EC of butylate at a rate of 15% lost over 95% of the ai in 48 hr in both the dry and the wet test.

TABLE 14

Effect of EPTC Formulations on Green Weight Yield of Robust Purple Foxtail Over Time

Harvest no.	No. days from initial treatment	1.68 kg/ha		3.36 kg/ha	
		Formulation[a]	Gr. wt (g)	Formulation	Gr. wt (g)
1st	28	Untreated	3.8 A[b]	Untreated	3.8 A
		EC-62	0 B	EC-62	0 B
		60	0 B	60	0 B
		61	0 B	61	0 B
		59	0 B	59	0 B
		58	0 B	58	0 B
2nd	56	Untreated	2.7 A	Untreated	2.7 A
		EC-62	1.4 B	EC-62	0.5 B
		60	0.8 C	60	0.5 B
		59	0.8 C	58	0.1 B
		58	0.5 C	61	0.1 B
		61	0.3 C	59	<0.1 B
3rd	86	Untreated	2.5 A	EC-62	2.6 A
		60	2.3 AB	Untreated	2.5 AB
		59	2.2 AB	58	2.2 AB
		58	2.2 AB	60	2.0 AB
		62	2.0 AB	61	2.0 AB
		61	1.6 B	59	1.7 B
4th	116	No significant differences within formulations at all concentrations			

[a] Formulation numbers refer to the text or Table 12. EC-62 code number for EC of EPTC.

[b] The values in the column within each concentration and in each harvest followed by the same letter are not significantly different according to Duncan's multiple range test at a 5% level.

and the other was a double-encapsulated product containing 20% latex polymer and 22% EPTC. In the greenhouse, the double-encapsulated product was compared with EPTC applied as the EC. The materials were applied to soils at a rate of 3 lb/acre and were incorporated into soils seeded with robust purple foxtail.

At 31 days after seeding, the plants were removed and weighed, and the pots re-seeded. This procedure was repeated at 52 and 87 days, and the final harvest was at 120 days. Results are shown in Table 18.

At 31 and 52 days, both the EC and the starch xanthide formulations gave excellent control of foxtail. However, between 52 and 87 days, the EC completely lost its effectiveness, whereas the starch product still gave good control. Between 87 and 120 days, the starch product lost its effectiveness. The greater weight of plant material in both treated pots over the control is due to the greater fertility level remaining in the soil of these pots.

In the field, the EC, single-encapsulated and double-encapsulated products were applied at active ingredient levels of 3 and 6 lb of ai per acre, incorporated, and over-seeded with yellow, giant, and giant green foxtail. Natural populations of pigweed, lambsquarter, smartweed, jimson, and velvetleaf were the dominant broadleaf weeds. Weed stand counts and weed weights by species were made on each plot 47 days after application. Results are shown in Table 19.

When applied at the 6 lb of ai per acre level, excellent control of all vegetation was obtained 105 days after treatment with the double-encapsulated formulations.

In the field of larvicides, a feed additive has been considered as a way to control

TABLE 15

Effect of Butylate Formulations on Green Weight Yield of Robust Purple Foxtail Over Time

Harvest no.	No. days from initial treatment	1.68 kg/ha		3.36 kg/ha	
		Formulation[a]	Gr. wt (g)	Formulation	Gr. wt (g)
1st	28	Untreated	3.5 A[b]	Untreated	3.5 A
		69	0.4 B	69	0.1B
		66	0.1 B	66	<0.1 B
		67	0.1 B	67	0 B
		68	<0.1 B	EC-70	0 B
		EC-70	<0.1 B	68	0 B
2nd	56	Untreated	2.9 A	Untreated	2.9 A
		EC-70	0.6 B	EC-70	0.1 B
		66	0.3 C	68	0.1 B
		67	0.2 C	66	0.1 B
		68	0.2 C	67	0 B
		69	0.2 C	69	0 B
3rd	86	Untreated	2.9 A	Untreated	2.9 A
		EC-70	1.3 B	EC-70	0.5 B
		68	0.8 BC	69	0.1 B
		66	0.3 C	66	0.1 B
		67	0.2 C	68	0.1 B
		69	0.2 C	67	0.1 B
4th	116	EC-70	4.AA	EC-70	4.4 A
		66	4.2 A	67	4.0 A
		67	3.9 A	69	4.0 A
		69	3.8A	68	3.8 A
		68	3.4 AB	66	3.8 A
		Untreated	2.4 B	Untreated	2.4 B

[a]   Formulation numbers refer to the text or Table 13. EC-70 code number for EC of butylate.

[b]   The values in the column within each concentration and in each harvest followed by the same letter are not significantly different according to Duncan's multiple range test at a 5% level.

TABLE 16

Percent Weed Control[a] by Delayed Incorporation of Butylate Formulations

Formulation	Rate	Incorporation	
		0 hr	24-hr delay
Starch xanthide-butylate[b]	3 lb ai/acre	80	75
Starch xanthide-butylate[c]	3 lb ai/acre	85	67
Butylate (EC)	3 lb ai/acre	87	10

[a]   Average control of barley, foxtail, watergrass, wild oat, crabgrass, annual ryegrass, and shattercane.

[b]   Starch xanthate of a D.S. of 0.3; $NaNO_2$ was used for crosslinking about 25% butylate.

[c]   Same as Footnote b except containing 35% butylate.

TABLE 17

Percent Weed Control by Delayed Incorporation of Butylate Formulations

		Incorporation	
Formulation	Rate	0 hr	24-hr delay
Starch xanthide-butylate[a]	4 lb ai/acre	77	68
Starch xanthide-buylate[b]	4 lb ai/acre	79	69
Butylate (EC)	4 lb ai/acre	82	28

[a] Starch xanthate of a D.S. of 0.3; $NaNO_2$ was used for crosslinking; 25.5% butylate.
[b] Same as Footnote a except larger particle size.

TABLE 18

Weed Control by Starch Xanthide-EPTC Formulation

	Yield (g/pot)			
Treatment	31 days	52 days	87 days	120 days
Untreated	1.6	2.0	2.8	2.2
EPTC (EC)	0	0.1	2.7	2.8
Starch xanthide-EPTC	0	0.01	0.4	3.9

TABLE 19

Weed Control by Single and Double Encapsulated EPTC Formulations

	Total weed	
Treatment[a]	Count	Weight (g)
Untreated	95.3	924.8
EC-EPTC	48.0	247.8
Starch xanthide-EPTC (single)	33.7	186.0
Starch xanthide-EPTC (double)	33.0	31.0

[a] Applied at a rate of EPTC of 3 lb/acre.

flies that breed in the droppings of hens housed in caged-layer poultry operations. Miller reviewed the status of this method and reported[23] that although many compounds (mainly organophosphorus insecticides) had been tested for such use, none was registered, primarily because residues of the insecticides were found in the eggs when the compounds were fed at the levels needed for fly control. Several formulations containing the larvicide diflubenzuron [N-[[(4-chlorophenyl)amino]carbonyl]-2,6-difluorobenzamide] (Dimilin®, TH 6040) were prepared[24] using starch xanthate as the encapsulating agent. The results are summarized in Table 20.

As can be seen from the table, residues of the larvicide in eggs using starch xanthate

TABLE 20

Effects of Feeding Hens Technical and Starch Xanthate Formulations of Diflubenzuron in the Ration at a Level of 5 ppm AI

AI in formulation (%)	Formulation identification	Inhibition of development (%)	Residues in eggs (ppm)
99.4	Technical[a]	94	0.58
15.0	Starch xanthate[b]	92	0.38
25.0	Starch xanthate[b]	92	0.30
46.0	Starch xanthate[b]	90	0.23
14.0	Starch xanthate[c]	96	0.20
48.0	Starch xanthate[c]	97	0.33
19.6	Starch xanthate[c]	83	0.37

[a] Air-milled, 1 to 5 μm.
[b] Starch xanthate of a D.S. of 0.17; resorcinol-formaldehyde as additive; $NaNO_2$ as oxidant.
[c] Starch xanthate of a D.S. of 0.35; latex SBR 1502 as additive; H_2O_2 as oxidant.

formulations varied between 0.20 and 0.38 ppm, whereas with the technical, the residues were 0.58 ppm. The diflubenzuron was mixed with the feed at a level of 5 ppm.

In another series, cellulose xanthate was used for encapsulation in place of starch xanthate. Then, the residues dropped to 0.1 ppm, even though the diflubenzuron was mixed with the feed at a level of 15 ppm.

VI. METHODS OF ENCAPSULATION

A. Starch Xanthate

For the purpose of encapsulation, corn starch (162 g, dry basis) is slurried in water (1 ℓ) and mixed with carbon disulfide (40 mℓ), followed by sodium hydroxide (40 g) dissolved in water (350 mℓ). Gelation occurs immediately. The mixing is continued until a homogeneous mixture is obtained. In about 1 hr, the xanthate is ready to be used. The product has a D.S. of 0.3 and is useful for encapsulation for up to 30 days when kept at 5°C. Nevertheless, to minimize the formation of byproducts, it is advisable to prepare a fresh batch as needed. For the preparation of starch xanthate with a D.S. of 0.2, the above recipe is followed, except that the amounts of carbon disulfide and sodium hydroxide are halved.

These xanthates contain 12 to 14% solids. Higher solids content (up to 20%) can be made, but the gel thus produced is too thick to handle easily. A significant reduction in viscosity occurs upon mixing the xanthate in a high-speed Waring Blender® for 10 to 20 sec. For certain formulations, one might use acid-modified starch or acid-modified flour (both are commercially available). The xanthates here are much less viscous, and a solids content of 50% or higher can be easily achieved.

Other polyols can also be used for encapsulation via xanthation. Polymers, such as cellulose and polyvinyl alcohol, are easily xanthated, although one might not apply the same recipe as with starch.

B. Encapsulation of Liquids Such as Butylate

1. Using H_2O_2 as oxidant, starch xanthate at a D.S. of 0.3 (200 g) is cooled to 5°C and mixed thoroughly with butylate (6.7 E, 10 g), followed by the addition of glacial acetic acid (8 mℓ) and H_2O_2, 30% (5.5 mℓ). After being mixed and allowed

to stand for 15 min, the mixture is filtered and pressed under a rubber dam to remove most of the water. The product is ground with a Waring Blender® and dried in a hood to yield 38 g of granular material containing 16.4% butylate.
2.   Using NaNO₂ as oxidant, starch xanthate at a D.S. of 0.17 (120 g) is mixed with 50% NaNO₂ solution (3.5 mℓ), followed by butylate (6.7 E, 10 g) and glacial acetic acid (10 mℓ). The product is isolated and dried, as above, to yield 25 g of product containing 22.4% butylate.

These procedures to encapsulate butylate are easily adaptable to encapsulate low-melting products and compounds with a high degree of solubility in organic solvents.

C. Encapsulation of Solids Such as Coumaphos
For encapsulation of this type of material, the finest powder (air milled or alike in mesh size) should be used.

Starch xanthate at a D.S. of 0.3 (1200 g) is cooled to 5°C and mixed in a Waring Blender® with 20 g of coumaphos (technical grade). Acidification and crosslinking are performed with the addition of glacial acetic acid (45 mℓ) and H_2O_2, 30% (25 mℓ). The product is isolated and dried, as above, to yield 193 g of encapsulated coumaphos containing 10.4% active ingredient.

VII. FUTURE LINES OF RESEARCH

The technology of encapsulation of biologically active ingredients within the starch matrix is still in its infancy. Nevertheless, it does show a good potential of acceptance because of its simplicity, its low cost, and mainly because it does improve the performance of important classes of pesticides. Still much fundamental and applied research is needed in this exciting field. Among questions to be answered are

1.   Does this type of encapsulation reduce the toxicity of pesticides? It was noted, for example, that encapsulated 2,4-D free acid or the ester analogue eliminates the odors related to this compound.
2.   What is the mechanism of release of active ingredients from the starch matrix? It appears that diffusion (and not biodegradation) throughout the matrix plays a dominant role.
3.   What should the properties of matrices be to efficiently entrap water-soluble compounds, such as fertilizers? Thus far, the starch matrix has proved to be of limited use. Other systems, especially hydrophobic, should be investigated.
4.   Which machinery is best suited to apply these granular encapsulated materials?
5.   What other starch derivatives could be used to substitute for xanthate? What properties would these new formulations have in terms of rate of release?

REFERENCES

1. **Gray, R. A. and Weierich, A. J.,** Factors affecting the vapor loss of N,N-dipropylthiocarbamate from soils, *Weeds,* 13, 141, 1965.
2. **Smith, A. E. and Verma, B. P.,** Weed control in nursery stock by controlled release of alachlor, *Weed Sci.,* 25, 175, 1977.
3. **Siegmann, A., Narkis, M., and Tirosh, N.,** Slow release of herbicides from polymeric granules, *Agric. Res. Organ. Volcani Central Division Sci. Publ.,* 82, 40, 1975.

4. **Allan G. G., Chopra, C. S., Neogi, A. N., and Wilkins, R. M.,** Controlled release pesticides. II. Synthesis of herbicide-forest solid waste combinations, *Tappi,* 54, 1293, 1971.

5. **Mehltretter, C. L., Roth, W. B., Weakley, F. B., McGuire, T. A., and Russell, C. R.,** Potential controlled release herbicides from 2,4-D-esters of starches, *Weed Sci.,* 22, 415, 1974.

6. **Shasha, B. S., Doane, W. M., and Russell, C. R.,** Starch encapsulated pesticides for slow release, *J. Polym. Sci.,* 14, 417, 1976.

7. For general references on starch see: **Kerr, R. W.,** Ed., *Chemistry and Industry of Starch,* 2nd ed., Academic Press, New York, 1950; **Radley, J. A.,** Ed., *Starch and Its Derivatives,* 4th ed., Chapman and Hall, London, 1968; **Whistler, R. L. and Paschall, E. F.,** Eds., *Starch: Chemistry and Technology,* Vol. 1 and 2, Academic Press, New York, 1965 and 1966; **Pigman, W. and Horten, D.,** Eds., *The Carbohydrates, Chemistry and Biochemistry,* Vol. 2b, Academic Press, New York, 1970.

8. **Rao, S. R.,** *Xanthates and Related Compounds,* Marcel Dekker, New York, 1971, 395.

9. **Hamerstrand, E., Carr, M., Hofreiter, B., and Russell, C. R.,** Starch xanthide in pilot paper machine trials, *Staerke,* 28, 240, 1976.

10. **Buchanan, R. A., Seckinger, H. L., Kwolek, W. F., Doane, W. M., and Russell, C. R.,** Controlling particle size and structure of starch xanthide in elastomer masterbatches, *J. Elastomers Plast.,* 8, 82, 1976.

11. **Abbott, T. P., James, C., Doane, W. M., and Russell, C. R.,** Injection molding of conventional formulations based on starch-encased powdered rubber, *J. Elastomers Plast.,* 7, 114, 1975.

12. **Wing, R. E., Doane, W. M., and Russell, C. R.,** Insoluble starch xanthate: use in heavy metal removal, *J. Appl. Polym. Sci.,* 19, 847, 1975.

13. **Rogers, C. E.,** Structural and chemical factors controlling the permeability of organic molecules through a polymer matrix, in *Controlled Release Pesticides,* ACS Symp. Ser. 53, Scher, H. B., Ed., American Chemical Society, Washington, D.C., 1977, chap. 2.

14. **Crank, J. and Parks, G. S.,** Eds., *Diffusion in Polymers,* Academic Press, New York, 1968.

15. **Rogers, C. G.,** Solubility and diffusivity, in *Physics and Chemistry of the Organic Solid State,* Vol. 2, Fox, C., Labes, M., and Weissberger, A., Eds., Interscience, New York, 1965, chap. 6.

16. **Machin, D. and Rogers, C. E.,** Free volume theories for penetrant diffusion in polymers, *Polym. Prepr. Am. Chem. Soc. Div. Polym. Chem.,* 11, 142, 1970.

17. **Hopfenberg, H. B.,** Ed., *Polymer Science and Technology,* Vol. 6, Plenum Press, New York, 1974.

18. **Crank, J.,** *The Mathematics of Diffusion,* 2nd ed., Clarendon Press, Oxford, 1975.

19. **Flynn, G. L., Yalkowsky, S. H., and Roseman, T. J.,** Mass transport phenomena and models. Theoretical concepts, *J. Pharm. Sci.,* 63, 479, 1974.

20. **Doane, W. M., Shasha, B. S., and Russell, C. R.,** Encapsulation of pesticides within a starch matrix, in *Controlled Release Pesticides,* ACS Symp. Ser. 53, Scher, H. B., Ed., American Chemical Society, Washington, D.C., 1977, chap. 7.

21. **Feldmesser, J., Shasha, B. S., and Doane, W. M.,** Nematicides in starch for controlled release, in *Proc. 1976 Int. Controlled Release Pesticide Symp.,* Cardarelli, N. F., Ed., University of Akron, Ohio, 1976, 6.18.

22. **Feldmesser, J. and Shasha, B. S.,** Evaluation of controlled-release DBCP-starch xanthide granules in laboratory and greenhouse soil tests, in *Proc. 1977 Int. Controlled Release Pesticide Symp.,* Goulding, R. L., Ed., Oregon State University,, Corvallis, 1977, 205.

23. **Miller, R. W.,** Larvacides for fly control, *Bull. Entomol. Soc. Am.,* 16, 154, 1970.

24. **Miller, R. W.,** Poultry feed additives for fly control: improvement in efficiency with controlled release formulations, in *Proc. 1977 Int. Controlled Release Pesticide Symp.,* Goulding, R. L., Ed., Oregon State University, Corvallis, 1977, 264.

25. **Schreiber, M.,** Efficacy and persistence of starch encapsulated EPTC, in *Proc. North Central Weed Control Conf.,* 31, 143, 1976.

Chapter 12

PINE KRAFT LIGNIN AS A PESTICIDE DELIVERY SYSTEM

H. T. DelliColli

TABLE OF CONTENTS

I. INTRODUCTION

Modern man is continually being challenged to develop new, safe, and more economical means of providing for his own well being. In many of these instances, the key to successfully meeting these challenges lies in the control of biological systems. Traditionally, this problem has been dealt with through the design of new biologically active materials, such as drugs and agricultural chemicals. Regardless of the specific nature of the chemical substance being considered, several factors grossly reduce the efficiency of the agent. Many of these factors are associated with the means of delivering the chemical to the target organism, be it insect, weed, or bacterium.

Conventional means of pesticide use are classic examples of delivery systems that reduce the effectiveness of the very agents they are designed to disseminate. Despite the problems associated with the use of pesticidal chemicals, their use cannot be discontinued without inviting cultural and economic disaster brought about by critical food shortages. In view of this, it is obvious that the development of more efficient and safer delivery systems for agricultural chemicals is both desirable and necessary.

Since the banning of DDT and other halogenated hydrocarbons by the Environmental Protection Agency, extensive use has been made of organophosphate and carbamate pesticides. These types of materials are much more susceptible to environmental degradation and much more toxic than the more persistent materials.

Factors such as chemical and biological degradation, runoff, leaching, and evaporation often require the use of excessively high doses of pesticide in order to maintain biological control. Compensation for these losses by higher dose rates and more frequent applications raises both material and labor costs to the user. The high toxicity of some chemicals also increases the risk to individuals associated with the dissemination of these materials.

In many instances, pesticidal chemicals have adverse effects on the very crops they are charged with protecting. The phytotoxicity associated with certain chemicals often manifests itself in the form of reduced crop yield because of plant damage.

In an effort to meet and satisfy demands for pesticide performance and to simultaneously overcome the above-mentioned shortcomings of the conventional dissemination system, the agricultural chemical industry, as well as agriculture itself, has looked towards new pesticide delivery systems. These systems must enable currently available chemicals to do the jobs assigned to materials not yet in the marketplace. Many of these delivery systems fall into the rather broad category of controlled release systems.

The number of such systems which have been developed is many and varied. They range in complexity from simple mineral-based granules and powders to more complicated polymeric devices and microcapsules and, finally, to hollow fibers and highly sophisticated osmotic pumps.

The following paragraphs describe the use of one such system, namely, pine kraft lignin.

II. LIGNIN

Lignin, the most abundant naturally occurring aromatic polymer in the world, constitutes a portion of the cell wall of most dry land plants. In fact, lignin, exceeded in abundance only by cellulose, is the primary nonpetroleum source of the aromatic nucleus in nature.

As it exists in the living tree, lignin, or, more appropriately, native lignin, serves several functions. It imparts structural rigidity to the stems of the plant by acting as a bonding agent between cells. It also plays a role in decreasing the permeation of water

TABLE 1

Functional Groups in Pine Kraft Lignin

Functionality	Mol/1000 g of lignin
Phenolic hydroxide	4.0
Aliphatic alcohol	2.7
Carboxyl	0.9
Carbonyl	0.8
Double bond	0.6
Methoxyl	5.0

across cell walls in the vascular system of the tree and imparts resistance to attack by certain microorganisms.

Lignin is produced in nature from the glucosides of coniferyl and sinapyl alcohols.

$$
\begin{array}{cc}
\gamma \quad CH_2OH & \gamma \quad CH_2OH \\
| & | \\
\beta \quad CH & \beta \quad CH \\
\| & \| \\
\alpha \quad CH & \alpha \quad CH
\end{array}
$$

Coniferyl Alcohol   Sinapyl Alcohol

The actual polymerization is preceeded by an enzymatic dehydrogenation, which leads to a free radical with electron density delocalized to the phenolic oxygen and carbons 5, 1, and β. According to Nimz[1], random coupling through each of these reactive sites leads to a material with a statistical distribution of at least ten different types of bonds.

III. KRAFT LIGNIN

Native lignin is not available for industrial purposes. The material to be discussed in the following text is kraft lignin, a byproduct of the kraft pulping process. Technical kraft lignin is a polyphenolic polyether soluble in alkaline aqueous solutions, such as the kraft pulping liquors. It is, however, insoluble in neutral and acidic media. This contrasts with the sulfite lignins, which are isolated from the liquors of the sulfite pulping process and which are soluble under neutral and moderately acidic conditions.

The organic chemistry of lignin is a complex area and is the subject of several volumes.[2,3,4] The principal functional groups of pine kraft lignin which are distributed statistically are described in Table 1.

The phenols are usually sterically hindered by one or more ortho methoxyl groups, a factor which accounts for the lack of toxicity of kraft lignin to higher organisms.

The physical chemistry of kraft lignin has at present not been elucidated, a situation due in part to the random and not yet fully understood mode of in vivo synthesis of the parent native lignin, the statistical distribution of its functionality, and the changes in structure and composition which take place during the pulping process.

Aside from solubility in alkaline solutions as well as polar organic solvents, such as DMSO and DMF, current available data[5,6,7] indicate that technical kraft lignin has an intrinsic viscosity, $[\eta] = 6$ ml/g. Other hydrodynamic data, such as diffusion and sedimentation coefficients, suggest that the dissolved macromolecule exhibits properties which would describe a roughly spherical hydrodynamic entity. Recently, however, Goring[8] has suggested that isolated lignins are flat, disc-like structures with a thickness of about 15Å. Average mol wt are 1600 and 1050 daltons for pine and hardwood lignins, respectively. Other molecular weight data indicate a moderate polydispersity, $\bar{M}w/\bar{M}n$, for lignins; 2.2 for pine and 2.8 for hardwood.

IV. KRAFT LIGNIN AS A PESTICIDE DELIVERY SYSTEM

When compared with carriers currently available to the pesticide industry, kraft lignin offers certain advantages. The highly aromatic nature of lignin makes it an excellent protective matrix for chemicals sensitive to degradative processes initiated by the UV radiation in sunlight. Pesticides of this type include the aniline derivatives and natural and synthetic pyrethroids. Secondly, the antioxidant properties of lignin add further stability to chemically unstable pesticides, a characteristic of many of the "non-persistent" agents in use today or being developed for future use.

Biodegradability has also been identified as a property of pine kraft lignin. In view of the current and, undoubtedly, future trends in chemical pest control, this property makes the above-mentioned polymer highly desirable as a carrier. In the soil, lignin is microbially degraded to precursors of humus by several microorganisms, such as the white rot fungi, which belong to the family, Basidiomycetes. Several of the pathways of lignin degradation by microflora into humus were reviewed and described by Christman and Oglesby.[9] Very recently, Lundquist et al.[10] demonstrated the degradability of kraft lignin, bleached kraft lignin, and water soluble lignin sulfonates by *Phanerochaete chrysosporium*. Lundquist concluded that the industrial lignins (kraft and bleached kraft) were bioalterable and under conditions favorable to the metabolic processes of the microorganisms, substantially biodegradable.

Easy conversion of a composite consisting of a lignin carrier and active material to a formulation readily usable by its intended user, the grower, is a property of lignin systems. For example, wettable powder and flowable (suspension concentrate) formulations have been prepared from simizine, atrazine, pentrachloronitrobenzene (PCNB), hexachlorophene, methyl parathion, and 2,4-D composites. Granules utilizing the same actives, as well as several other phenoxys, substituted benzoic acids, and carbamates, have been prepared. All composites utilized technical kraft lignin, Indulin AT®, as a carrier. These types of formulations fit readily into current cultural practices.

On the other hand, microcapsules find almost exclusive use in flowable systems, while synthetic polymeric matrix-based composites fit into granular formulations or special devices, such as strips and animal collars.

When several composites of technical kraft lignin and organophosphates, such as Ethoprop® (Mobil Chemical), and carbamates, such as Bux® (Chevron), took the form of viscous gums or deformable solids which could only be ground cryogenically, it became apparent that Indulin AT® was not universally acceptable as a carrier. Even after the low temperature grind, the formulations had absolutely no shelf stability. Both small particle formulations and granules either sintered within hours at temperatures as low as 30°C or, in many cases, actually liquified within days at slightly higher temperatures (<50°C). These observations led to the development of a chemically modified lignin carrier prepared by modifying the parent technical kraft lignin to produce a reversably swellable porous matrix. With few exceptions, this carrier made possible the preparation of stable formulations of those agricultural chemicals which plasticized the parent lignin.

In another instance, the lignin was converted to a second modification. The carrier system produced was again swellable, but underwent an 11-fold volume increase upon hydration compared to a two-fold change displayed by the previous carrier. This carrier was found useful for water-soluble pesticides, such as the amine salts of 2,4-D, etc.

The above modifications were made possible because of one of the major advantages of kraft lignin. This factor is its ability to be chemically modified. Work to be discussed in the ensuing paragraphs describes the field performance of these composites, each of which employs one of the basic carriers described above. These experimental systems are kraft lignin/Terraclor® (PCNB) and modified lignin/2,4-D (dimethyl amine salt).

V. CARRIER SELECTIONS

As reported earlier,[11] the choice of carrier systems for any given technical pesticide is still largely empirical. Some insight into the extent to which the carrier and active ingredient interact can be gained by calculating the solubility parameter of the pesticide. Generally, estimation of this parameter with the structural group considerations of Small[12] will suffice. Experience has shown that where δ for the pesticide falls between 10.5 and 12 $(cal/cm^3)^{1/2}$ the likelihood of interaction is high. A value of $\delta = 11.5$ is used for technical kraft lignin.

VI. LIGNIN/2,4-D FIELD RESULTS

A potential controlled release system employing 2,4-D, as the dimethyl amine salt, was tested under field conditions. The formulation, which consisted of 25×40 mesh granules with 10% active ingredient, was prepared by immersing the dehydrated carrier in a dilute solution of the 2,4-D salt. Swelling of the carrier took place within minutes, thereby incorporating all of the amine solution within the interior and on the surface of the lignin.

The windward side of a ditch bank was chosen as the test site. A series of ten trays with 1 ft² openings were filled with presterilized soil and treated with lignin/2,4-D at the rate of 0.5 lb of active ingredient per acre. An alternate series of ten trays were treated with a commercial 4-lb/gal DMA formulation of 2,4-D at the same rate (0.5 lb a.i./acre), while a third ten-tray arrangement, which served as a control, received no herbicide treatment. It was intended that these trays catch windborne seed and allow it to germinate where no herbicide was employed or prevent germination where chemical was present. Use of this arrangement eliminated the possibility of preexisting root systems or seeds in the test area.

The test plots were examined every 4 weeks, and all growing plants were removed and counted. As a further check on the efficiency of the herbicide system, five bean seeds of the Kentucky Wonder variety were planted in each compartment. At each 4 week interval all bean seeds or plants were also removed and counted.

Data are reported as percent suppression of vegetation relative to the total stem count on the control plots. Larger numbers indicate greater weed control. Performance of the test systems against weeds of several unknown species is reported in Table 2.

These data are also represented graphically in Figure 1. Both representations of the data indicate similar initial biological activities for both systems. However, the level of suppression of the conventional system began to fall off rapidly and represented approximately 40% suppression 4 weeks after application. The level of control displayed by the lignin-based system remained at the 70 to 90% level of suppression until

TABLE 2

Suppression of Weeds by Conventional and Controlled Release 2,4-D

Elapsed time since herbicide application (Weeks)	Suppression (%)		No. of stems on control plots
	Controlled release	Conventional	
0	64	68	74
4	79	37	96
8	86	14	133
12	89	−2	237
16	69	3	304
20	68	2	276
24	9	−13	300

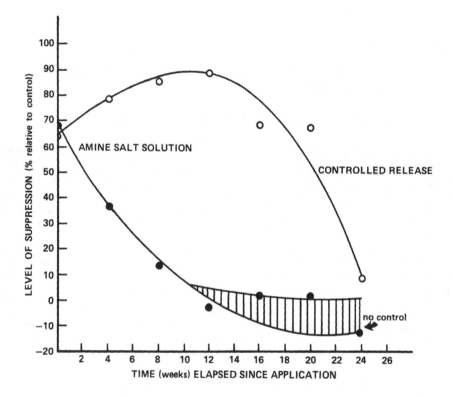

FIGURE 1.   Weed suppression by conventional and controlled release 2,4-D.

some time between 20 and 24 weeks after application. Approximately 10% control was achieved by the commercial system at week 8, with no control evident after this period. Negative suppression values are assumed to represent no control of vegetation. At several instances during the evaluation period, grasses unaffected by 2,4-D were planted in plots of each series. Germination of this seed was taken as proof of the viability of the soil matrices.

Table 3 represents the data describing the performance of the test systems against the bean plants and/or seeds. Graphical representation of these data appears on Figure 2. Examination of the data again points to an apparent breakdown or loss of activity

TABLE 3

Appearance of Bean Sprouts in Plots Treated with Conventional
and Controlled Release 2,4-D

Time elapsed since application (Weeks)	Sprouts breaking ground (%)		
	Controlled release	Conventional	Control
0	0	0	96
4	0	0	90
8	0	62	100
12	0	100	100
16	0	96	98
20	52	100	98
24	100	100	100

of the conventional system between weeks 4 and 8 and weeks 16 and 20 for the controlled release formulation. Performance of the lignin formulations reveals that one of the classical symptoms of controlled release, i.e., extended periods of effective biological control, is possible with lignin.

VII. LIGNIN/PCNB FIELD RESULTS

Aside from extended periods of control or a reduction in the number of applications of a given pesticide per growing season, controlled release can be used to ameliorate the performance of certain chemicals. For example, the fungicide, pentachloronitrobenzene (PCNB), known to be effective against the fungi, *Rhizocitonia solani* and *Sclerotinia sclerotiorum*, is also extremely phytotoxic and frequently traumatizes the plants with which it comes in contact. This chemically induced injury frequently results in yield loss.

In this particular instance, a composite system consisting of 55% PCNB and 45% lignin was converted to a 50% wettable powder.

The purpose of this test was the establishment of the biological efficacy of the lignin-based composite under "commercial" growing conditions. Fungicidal material was applied by farm workers employed by the grower; the plots were tended in a manner identical to commercial acreage immediately adjacent to the test site; and harvest was performed by migrant farm labor.

The test site was subdivided into four plots of approximately 0.25 acres each and planted with tomatoes of the Walter variety in a density of 4200 plants per acre.

The application of fungicide took place when the plants had been in the field for 14 days. The plots were cultivated less than 60 min after application and two additional times before harvest. No additional fungicides were applied to the test area after the initial dose. Maneb (manganese ethylenebisdithiocarbamate) was applied weekly during the season to control airborne microorganisms on the commercial acreage, but not on the test areas. Within 5 days of harvest, the use of Maneb was halted, and Bravo was used during the harvest period itself, which took place on June 27, 1977. Prior to the actual harvest, several pertinent observations were made. The check plot which received no PCNB showed large numbers of lesioned fruit and areas where entire plants were either dead or dying. Similar observations were made at another site on plots deliberately infected with *R. solani*. In fact, lesions were observed in most cases where fruit made contact with the ground and where soil was thrown onto the plants and fruit by cultivation equipment or by spray rigs applying insecticides to the plots.

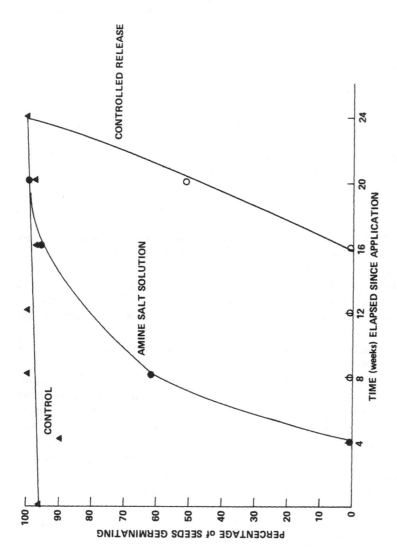

FIGURE 2.   Development of sprouts in plots treated with conventional and controlled release 2,4-D.

TABLE 4

Field Results — Edisto Island, S.C., 1977; Lignin/PCNB

Treatment	Rate (lbs of active ingredient per acre)	Total no. of fruit	Acceptable fruit (%)	Reject fruit (%)	Total no. of acceptable fruit	Increase (%)
Lignin/ PCNB	20	471	90.1	9.9	424	
Terraclor® (PCNB)	20	241	85.5	14.5	206	106
Staked check	0	320	75.9	21.1	243	74
Staked check	0	320	88.1	11.9	282	50

Commercial Harvest Results

Treatment	No. of 1000-lb Pallet Boxes per acre
Lignin/PCNB	28.5
Terraclor® 75	17
Staked check	20.3
Unstaked check	17.3

On the other hand, little evidence of fruit-rot lesions was seen on the plots treated with the lignin/PCNB composite. Fruit virtually half buried in the soil was found to be free of any evidence of fruit rot.

Efficacy of the treatment was determined in two ways. First, ten plants were removed from the interior of each plot to eliminate edge effects, and the total number of fruit on each plant was removed, counted, and marked as either commercially acceptable or reject. Subsequently, all commercially acceptable fruit was picked and placed in 1000-lb pallet boxes.

The appropriate data are presented on Table 4. In this case, the apparent level of biological control was about the same as that displayed by the conventional wettable powder. This is indicated by the nearly equal values for the percentage of acceptable fruit harvested from the plots treated with the lignin-based and the conventional formulations.

Nonetheless, the advantage of the unconventional formulation is clearly spelled out in a yield advantage.

VIII. CONCLUSION

Pine kraft lignin has been shown to be effective as a pesticide delivery system. An extension of the active lifetime of an herbicide and the reduction of the phytotoxicity of a fungicide by lignin-based carriers have been demonstrated.

The results in the preceding paragraphs point to lignin as a pesticide delivery system which is capable of producing results indicative of controlled release formulations. These formulations are produced by currently available equipment and can be used in the same types of dissemination systems as the currently available pesticide formulations.

REFERENCES

1. **Nimz, H.,** Beech lignin - proposal of a constitutional scheme, *Angew, Chem. Int. Ed. Engl.,* 13, 313 1974.
2. **Sarkanen, K. V. and Ludwig, C. H.,** *Lignins, Occurrence, Formation, Structure, and Reactions,* Interscience, New York, 1971.
3. **Brauns, F. E. and Brauns, D. A.,** *The Chemistry of Lignin: Supplemental Volume Covering Literature for 1949—1958,* Acaemic Press, New York, 1965.
4. **Pearl, S. A.,** *The Chemistry of Lignin,* Marcel Dekker, New York, 1967.
5. **Lindberg, J. J., Tylli, H., and Magani, C.,** Notes on the molecular weight and the fractionation of lignin with organic solvents, *Pap. Puu,* 46, 521, 1964.
6. **Rezanowich, A., Yean, W. Q., and Goring, D. A. I.,** The molecular properties of milled wood and dioxane lignin, refractive index increment and ultraviolet absorption, *Sven. Papperstidn.,* 66, 141, 1963.
7. **Gupta, P. R. and Goring, D. A. I.,** Physicochemical studies of alkali lignins. III. Size and shape of the macromolecule, *Can. J. Chem.,* 38, 270, 1960.
8. **Goring, D. A. I.,** A speculative picture of the delignification process, in *Cellulose Chemistry and Technology,* A.C.S. Symp. Series. No. 48, Arthur, J. C., Ed., American Chemical Society, Washington, D.C., 1977.
9. **Christman, R. F. and Oglesby, R. T.,** Microbiological degradation and the formation of humus, in *Lignins, Occurrence, Formation, Structure, and Reactions,* Interscience, New York, 1971.
10. **Lundquist, K., Kirk, T. K., and Connors, W. J.,** Fungal degradation of kraft lignin, *Arch. Microbiol.,* 112, 291, 1977.
11. **DelliColli, H. T.,** Controlled release of pesticide from kraft lignin carriers, in *Controlled Release Pesticides,* A.C.S. Symp. Ser. No. 53, Scher, H. B., Ed., American Chemical Society, Washington, D.C., 1977.
12. **Small, P. A.,** Some factors affecting the solubility of polymers, *J. Appl. Chem.,* 3, 71, 1953.

Chapter 13

OTHER CONTROLLED RELEASE TECHNOLOGIES AND APPLICATIONS

Agis F. Kydonieus

TABLE OF CONTENTS

I. INTRODUCTION

This chapter comprises two separate sections: the first section pertains to controlled release (CR) technologies not discussed in detail in the previous chapters of this volume, and the second section summarizes the CR methods and applications uncovered in a search of the U.S. patent literature for 1977 and part of 1978.

The purpose of the first section is to bring to the attention of the reader the existence of these technologies, describe briefly their special characteristics, and, in general, complete the available information on CR methods.

The purpose of the second section is to uncover the latest technologies and applications in the patent literature, hopefully to provide an insight into future directions of the CR industry in terms of technology and application.

In both sections, the information has been grouped within the five major categories of CR technologies, i.e., (1) reservoir systems with a rate-controlling membrane, (2) reservoir systems without a rate-controlling membrane, (3) monolithic systems and laminated structures (4) other physical methods, and (5) retrograde chemical reaction systems.

II. OTHER CR TECHNOLOGIES

Most of the technologies discussed below, although unique in one way or another, are not completely novel and independent, but are extensions of the basic technologies discussed in the previous chapters.

A. Reservoir Systems with a Rate-Controlling Membrane

1. Organic Phase Separation

This process is, in effect, the inverse of the Aqueous Phase Separation (Volume II, Chapter 4) method and is used to microencapsulate hydrophilic substances. The polymeric microcapsules are produced by first preparing a dispersion of the hydrophilic solution in a solution of the hydrophobic encapsulating polymer in a nonaqueous, nonpolar solvent. The hydrophilic solution comprises the dispersed phase and exists in a form of tiny droplets, with the hydrophobic polymer solution constituting the continuous outer phase. When a second hydrophobic liquid, miscible with the polymer solution, but in which the polymer is itself insoluble, is added, the polymer precipitates around the droplets of the hydrophilic solution to form the polymeric microcapsules.[1] Microcapsules containing solutions of urea, citric acid, sodium acetate trihydrate, and other hydrophilic substances in water, ethylene glycol, or polyethylene glycol have been fabricated in this way. The polymer of choice appears to be ethyl cellulose.

2. Meltable Dispersions

This method is similar to the Organic Phase Separation method in that it is used to encapsulate or microencapsulate hydrophilic materials. Unlike the Organic Phase Separation process, the wall material is a low-melting waxlike substance. The capsules are produced by introducing the hydrophilic nucleus material and the waxlike wall material into a hot oily vehicle to form a three-phase system. When the waxlike material is heated and agitated, it liquefies, disperses in the oily hydrophobic phase, and coats the hydrophilic particles. Upon cooling, the walls solidify, and the encapsulation process is complete.[2] Magnesium hydride, ammonium nitrate, aspirin, stannous fluoride, toluene diisocyanate, and several other oil-insoluble substances have been encapsulated with this process. The polymers of choice include paraffin, tristearin, polyethylene, and diglyceride.

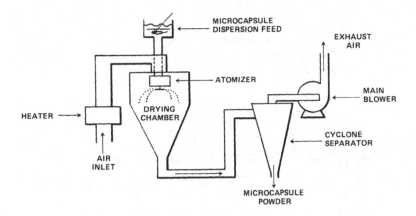

FIGURE 1. Schematic diagram of spray drier.

3. Spray Drying and Spray Congealing

These processes are similar to the Fluidized-Bed spray coating process described in Volume II, Chapter 7. Spray drying involves the dispersion of the core material in a coating solution and atomizing the dispersion into hot air. This rapidly removes the solvent from the coating solution, which solidifies the coating, providing a microcapsule product that is a free-flowing dry powder.[3] In spray congealing, no solvent is present; instead, the coating is a polymer melt, such as a wax, and the temperature drop in the spray drier is sufficient to congeal the mass and form the microcapsules.

These processes are usually conducted in various spray-drying equipment which are readily available in the marketplace.

The advantage of spray drying is the almost instantaneous means of producing a dry powder from a solution or a slurry. This is accomplished by reducing the liquid to a fine spray and mixing it with a hot gas. The powder is removed, and the gas containing the liquid from the solution or slurry in the form of moisture is exhausted into the atmosphere. The whole operation takes place in 5 to 30 sec.

Figure 1 shows a schematic diagram of a spray dryer. The main blower draws air through the system creating a negative pressure which prevents the escape of dust and toxic gases into the plant. An atomizer device (spray nozzle or centrifugal atomizer) disperses the liquid feed into a spray of droplets. Hot gases contact the spray droplets and evaporate moisture from the individual droplets.

Starting at the heater, air is raised to the desired drying inlet temperature and then mixes with the feed as it passes in the drying chamber; in evaporating the required moisture, the air is cooled to the outlet temperature. At the cyclone separator, the product is removed at the bottom while the air and moisture are exhausted.[4]

4. In Situ Interfacial-Condensation-Polymerization

This process is similar to the interfacial polymerization method discussed in Volume II, Chapter 6. The process comprises the dispersion of an organic phase containing isocyanate monomers into an aqueous phase. Upon heating the dispersion to an elevated temperature, the isocyanate monomers are hydrolyzed at the interface to form amines which, in turn, react with the unhydrolyzed isocyanate monomers to form the polyurea microcapsule wall.[5,6] The respective reactions are shown below:

$$-N = C = O + H_2O \xrightarrow{\text{heat}} -NH_2 + CO_2$$
$$\text{isocyanate} \qquad\qquad \text{amine}$$

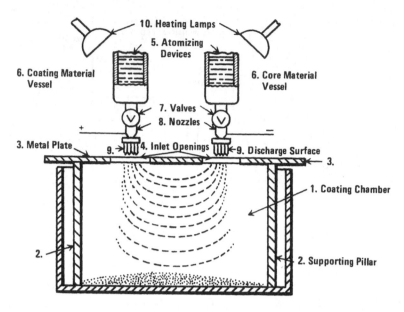

FIGURE 2.   Schematic diagram of electrostatic microencapsulation apparatus.

$$-N = C = O \ + \ \ -NH_2 \longrightarrow -\underset{\text{polyurea}}{\overset{\displaystyle H \quad O \quad H}{\underset{\displaystyle |}{\overset{\displaystyle |}{N}} - \underset{\displaystyle \parallel}{\overset{\displaystyle \parallel}{C}} - \underset{\displaystyle |}{\overset{\displaystyle |}{N}}} -$$

The isocyanate monomers used in the process are polymethylene polyphenyl isocyanate (PAPI) and toluene diisocyanate (TDI). The release rate of this microcapsule system can be regulated by varying the wall thickness, which is done by changing the weight percent of the isocyanate monomer in the organic phase and/or by altering the wall permeability, which can be varied by changing the ratio of PAPI to TDI. This ratio controls the crosslink density of the polyurea produced.

5. Interfacial Addition Polymerization

This process is very similar to the one described in *In Situ Interfacial-Condensation-Polymerization*. In this process, a selected monomer that can be polymerized into a solid is added to a hydrophobic liquid containing the active chemical to be encapsulated. The solution is then dispersed into a polar solvent, and polymerization is initiated by contact with an addition polymerization catalyst, which can be incorporated into the dispersion or the hydrophobic liquid. A preferred procedure utilizes a mixture of styrene and divinyl benzene in a 10:1 weight ratio and the catalyst, benzoyl peroxide. The process forms discrete, pressure-rupturable capsules which can contain dyes. The process is specifically designed for the coating of sheets to prepare pressure-sensitive copying papers.[7]

6. Electrostatic Encapsulation

This is a physical method, as is the Multiorifice Centrifugal process described in Volume II, Chapter 8. The electrostatic process encapsulates a liquid or solid core material while in the form of an aerosol or fine mist that bears an electric charge by envelopment of the core particle with a coating of an encapsulating substance, also maintained in the form of an aerosol or fine mist, but bearing an opposite charge.[8] The method involves electrostatically atomizing the core and coating material from separate sources as shown in Figure 2.[9] The core material has a higher surface tension than the coating material; thus, the former forms a negatively charged aerosol and the latter a positively charged aerosol. The two aerosols are fed into a mixing zone where they interact and achieve a state of substantial electrical neutrality, with formation of the encapsulated particles.

In the above-described process, the coating material can be a molten substance that can be sprayed, such as phthalate polyester or microcrystalline wax, or it can be synthetic resins, such as polyamides, polyvinyl esters, ethers, polystyrene, and acrylics dissolved in a solvent of high volatility.

The encapsulation process can also take place by condensation polymerization. This method necessitates the selection of two resin-forming reagents that will react rapidly to form a solid polymeric condensation product. One reagent is dissolved in the core material and the other in a suitable solvent. A polyamide capsule can thus be formed by reaction at the interface of a core material containing glycerine and hexamethylene diamine and a coating material containing toluene and adipyl chloride. The interfacial polymerization reaction results in the formation of polyhexamethylene diamine adipate.

The electrostatic encapsulation process is rather complex, and certain criteria must be met for the process to be feasible. For example, the core liquid must possess high surface tension such that the coating material will spread rapidly over its surface. In addition, for good atomization, the conductivity of the liquids should be between 2×10^{-3} and 5.8×10^{-6} ohm-cm. The basic requirements for the coating material include low surface tension and good wetting properties. Furthermore, the conductivity of the coating material should correspond to that of the core liquid. Several electrostatic encapsulation equipment designs are available.[8,10,11]

7. Pan Coating

This is used by the pharmaceutical industry to apply coatings to tablets and to prepare special granular formulations. The process can therefore be used to encapsulate numerous solids with film forming polymers.

Coating pans, sometimes called mushroom mixers, are one means of preparing glomules, tiny rolled-up balls of fine powders. The solid particles are rolled in the pan, and the liquid containing the dissolved polymeric film former is added by pouring or spraying.[12] The motion of the particles rubbing against each other helps to distribute the liquid evenly over the surface of each particle. The coating solvent is removed by evaporation, which may be speeded by passing hot air over the tumbling particles.

Coating pans are ordinarily held and rotated at the end of a shaft inclined at an angle of about 60° from the horizontal. They are usually made to rotate at 10 to 25 r/min and are available with smooth inside surfaces or with ribs or flights.

B. Reservoir Systems without a Rate-Controlling Membrane

1. Hydrogels

Hydrogels are a broad class of polymeric materials that swell in water as much as 30 to 90%, but do not dissolve therein. Hydrogels may contain entrapped within them such active chemicals as antibiotics, antibacterial agents, contraceptives, anticancer drugs, and even macromolecules such as proteins.[13-17] The active chemicals migrate through the imbibed water and are released to the environment of the hydrogel device. Owing to the high amounts of water absorbed, hydrogels exhibit good biocompatibility and are used in many biomedical applications, including CR therapeutic implants.

Hydrogels are produced by the polymerization of hydrophilic monomers or by chemical modification of existing polymers.[18,19] In both cases, adequate numbers of hydrophilic groups must be available to allow swelling, but no dissolution, in water.

Since highly swollen hydrogels are mechanically weak, the hydrophilic monomers are usually copolymerized with a hydrophobic monomer for control of the mechanical strength and the degree of swelling.

Hydrophilic monomers used in the preparation of hydrogels include: hydroxymethyl methacrylate, hydroxyethyl methacrylate (Hydron), methylene-bis-acrylamide, pyridine, sodium styrene sulfonate, crotonic acid, acrylamide, N-vinylpyrrolidone, 2,4-pentadiene-1-ol, and propylene glycol methacrylate.

In addition to biocompatability, the ability of hydrogels to release macromolecules in a controlled way is a significant achievement. It was recently shown that a copolymer of Hydron was capable in releasing insulin and the protein, soybean trypsin inhibitor, with a mol wt of 21,000.[14] This important property of hydrogels is probably due to the large pores or open structure of the swollen polymer imbibing normal bulk water. Studies on polyhydroxyethyl methacrylate hydrogels indicate that over 50% of the total hydrogel water is imbibed as normal bulk water.[20,21] Smaller amounts are bound water around hydrophylic groups and interface water structured around hydrophobic groups.

C. Monolithic Systems
1. Gels
Systems that consist of media with dissolved or dispersed particles, ranging from approximately 1 nm to a few microns, are called colloids. There are three classes of colloids, categorized by the manner in which the particles arise. They are (1) colloidal dispersions, (2) solutions of macromolecules, and (3) association colloids. The first two classes will be discussed because only they are used in producing gels to control the release of active ingredients.

a. Colloidal Dispersions
Unlike true solutions, which are homogeneous, colloidal dispersions are considered to be heterogeneous, with the dispersion medium (in which the dispersion takes place) as one phase and the substance dispersed, called the disperse substance, as the other phase. Gels are jellylike materials formed from the precipitation of sols, which are fluid colloidal solutions. Hydrophobic sols are dispersions in which there is very little attraction between the disperse substance and the medium, as occurs with various metals and salts in an aqueous medium. Hydrophilic sols are dispersions where there is a definite affinity between the medium and the disperse substance, and extensive solvation of the colloidal particles takes place. Examples of hydrophilic sols are dispersions of the hydrophobic type to which have been added, for example, agar-agar, carrageenan, gelatin, starch, or casein.

The coagulation and precipitation of sols can be effected by (1) the addition of certain ions to neutralize the electric charges, (2) boiling, (3) freezing, and (4) the addition of solvents, such as alcohol or acetone, together with small amounts of an electrolyte. The precipitate formed may or may not be gelatinous. However, if the conditions are right, it is possible to obtain the dispersed phase as a more-or-less rigid mass enclosing all of the liquid within it. The product is then called a gel, and the process is called gelation.

Depending on their nature, gels may be prepared, generally, by one of the following three methods: (1) cooling, (2) double decomposition or metathesis, or (3) a change of solvents. Gels of agar-agar, gelatin, and other substances of this kind are prepared readily by cooling a relatively concentrated dispersion of these substances in hot water. As the sol cools, the highly hydrated dispersed particles agglomerate into larger masses and eventually mat together to form a semirigid gel structure that entraps any free medium. The second method is illustrated by the formation of silicic acid gels on the addition of an acid to an aqueous solution of sodium silicate. The free silicic acid thus liberated is highly gelatinous due to hydration and sets more or less rapidly to a solid gel. Finally, some gels are formed by rather suddenly changing the solvent in which a substance is dissolved to one in which it is insoluble. An example of this is a gel of calcium acetate which is formed when alcohol is added rapidly to a solution of calcium acetate in water; the salt is suddenly thrown out of solution as a colloidal dispersion which subsequently sets to a gel containing all of the liquid.[22]

b. Solutions of Macromolecules

Polymers dissolve to form true solutions, as do solutes of lower molecular weight. However, because of the large size of the polymer molecules, these solutions behave as colloids. Many synthetic polymers, e.g., polystyrene, vinyl, acrylics, and nylon are soluble in several solvents, as are naturally occurring polymers, such as proteins, polysaccharides, gums, etc. It is possible in many instances to crosslink the chains of such polymers in solution to form a three-dimensional network containing all of the liquid and thus produce a gel. Examples of such gels include those formed by a sodium carboxymethyl cellulose solution in water precipitated by chromic potassium sulfate and those of potassium carboxymethyl cellulose and a trivalent metal cation, such as aluminum or chromic ion.[23,24]

c. Controlled Release from Gels

Gels are used almost exclusively for controlled release air freshener systems for the consumer market. The annual dollar volume at the retail level is estimated at 80 million dollars with such well known and established brand names as Glade®, Wizard®, and Airwick®.

The above-mentioned gels are prepared using carrageenan as the gelling agent, with smaller amounts of locust bean gum, agar-agar, and other water-soluble polymers also being used. The total amount of the gelling agent is 1.5 to 2.5% of the total weight of the gel. The dispersion medium is mainly water, between 85 and 95%, with the remainder being ethyl alcohol, ethylene glycol, a surfactant, a dye, and 2 to 4% perfume. Basic formulations can be found in the patent literature.[25] The gels are prepared by dispersing the carrageenan and other ingredients in the medium, heating the solution to 180°F to effect dissolution, and then cooling to room temperature to provide the air freshener gel.

Recently, gels have been prepared using the crosslinking of macromolecules in solution, as discussed above.[24] The formulation is similar to the one mentioned above except that the gelling agent is 3% sodium carboxymethyl cellulose crosslinked with basic chromic sulfate. The major advantages of these gels are that gelling is instantaneous, and no heating and cooling are required.

The major problems associated with gels are poor mechanical strength and some shrinkage of gels on standing, with an attendant exudation of water to the gel surface. The latter process, called syneresis, imparts to the gel a wet, unpleasant visual appearance. For these reasons, gels are marketed in plastic containers, which provide support to the gel, hide the appearance of syneresis, and present a pleasant-looking package to the consumer. As will be discussed in a later section the plastic container can be used as a secondary means for controlling release of the contents of the gel.

It is rather difficult to classify the gels under any of the CR classifications discussed in the first chapter of Volume I. One may consider gels as monolithic devices; however, gels differ from other devices of this type by the substantial change in the size of the gel itself as its contents are released. In this respect, the gel can be considered an ero-

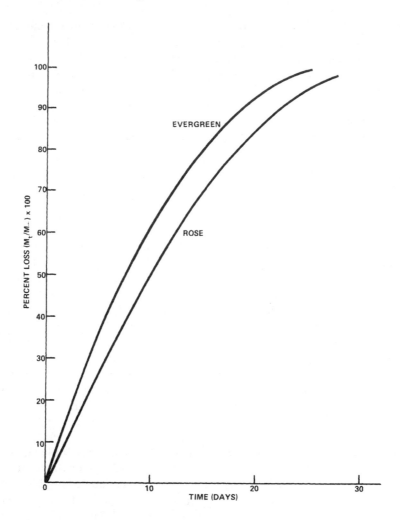

FIGURE 3.   Weight loss from air freshener gels.

dible device with volatilization being the mechanism of erosion. Again, this differs from the erodible devices discussed in Volume II, Chapter 1 by (1) the fact that carrageenan and other water-soluble polymers do not erode, leaving a solid polymeric matrix behind when the gel contents have been released and (2) the composition of the gel changes continuously with the percent carrageenan increasing and the percent of the volatile components decreasing.

A mathematical treatment taking into account the continuous change in the size and composition of the gel would be very interesting and novel. To the author's knowledge, such a mathematical treatment has not appeared.

Despite the complexities mentioned above, the air freshener gels release their contents in a substantially linear fashion. Figure 3 shows the cummulative percent loss from two commercially available air freshener gels containing evergreen and rose fragrance.[26] M_t represents the mass of agent released and $M\infty$ the mass of active agent initially available in the gel. The percent loss was based on the total volatile contents ($M\infty$) which were 160.4 and 165.4 g, respectively, for the evergreen and rose gels.

In accordance with equation 1.5 and section 3a of Volume I, Chapter 1, a plot of $M_t/M\infty$ vs. the square root of time should be linear (for the range $0 < M_t/M\infty < 0.6$)

if the gels were monolithic devices. Rearrangement of the data in Figure 3 shows that the relationship mentioned above does not hold true.

2. Other Monolithic Methods
a. The MGK® Matrix

This has been investigated by the McLaughlin Gromley King Company as a vapor delivery system.[27] Essentially, this invention consists of a mixture of polyethylene paraffin and a granulated, partially dehydrated metal salt properly prepared and admixed with a volatile active agent (essential oils, volatile pesticides, etc.).

Upon molding or granulation followed by compression molding, a block or other desired shape is formed that provides a relatively straight-line release of the vapor material.

A typical preparation from the patent illustrates the system:

17.5 g of low-molecular-weight paraffin having a melting point between 40 and 90°C are heated to 100°C. The melted paraffin is then heated to 125°C, and 17.5 g of small pellets of Epolene-E (polyethylene wax having a mol wt 1400 to 2500) are slowly added to the melted paraffin with stirring to disperse the Epolene-E uniformly throughout the melted paraffin. 50 g of granulated gypsum of a size between 100 and 400 mesh are added to the paraffin and Epolene-E with stirring. The mixture of gypsum, paraffin, and Epolene-E is heated to 135°C and when about one half of the water of hydration has been lost, which is indicated by bubbling, the mixture is cooled to 110°C. 15 g of odorant (3,5,5-trimethyl hexanal) are added to the mixture with stirring to disperse the odorant uniformly throughout the mixture. The mixture is then cooled rapidly to solidify the paraffin and Epolene-E. The solidified composition is then crushed into small particles. 5 g of Formalin are added to the mixture with stirring. The small particles with the Formalin are pressed into a mold, and the molded cohesive mass is then placed in a confined space, such as a room, to allow the odorant to diffuse to mask undesirable odors, such as tobacco smoke or animal odors.

Similarly, a volatile pesticide, such as DDVP (2,2-dichlorovinyl dimethyl phosphate), can be incorporated at 15 to 20% by weight to provide a pest control strip.

b. The Killmaster®

This approach is another method of preparing monolithic systems. Killmaster® is a lacquer containing 1% of the organophosphorous insecticide, chlorpyrifos, 2.1% polyurethane, and 96.9% petroleum distillates and other solvents. The lacquer is applied onto baseboards and around door and window casings in the home for control of cockroach populations. After application, the solvent evaporates leaving behind a monolithic polyurethane layer containing the chlorpyrifos. Tests in six houses using the Killmaster® controlled release formulation gave a 96.4% reduction in cockroach population 90 days after treatment.[28]

D. Other Physical Methods
1. Mechanical Dispenser

Mechanical dispensers are used as a secondary rate-controlling device, usually in conjunction with a primary CR device. Mechanical dispensers are decorative plastic containers with a threaded or other arrangement that enables one to close or open the container to a desired degree and thus regulate the release of the active ingredients.

As mentioned earlier, the mechanical dispensers are used in conjunction with primary CR devices. Some commercial products include (1) Airwick® Stick-ups, with the primary CR device being a fibrous mat impregnated with approximately 1 g of

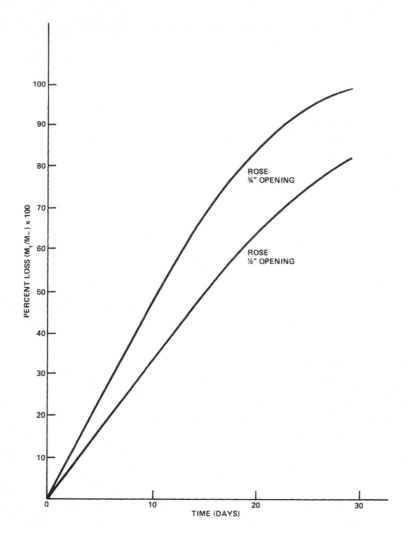

FIGURE 4.   Effect of mechanical dispenser opening on the weight loss from air freshener gels.

fragrance, (2) Wizard®, with the primary CR device being a carrageenan gel with approximately 4 g of fragrance, and (3) Raid® Solid, with the primary CR device being polyvinyl chloride monolithic device containing 20 g of the organophosphorus insecticide, DDVP, and related compounds.

All of the systems mentioned above have provisions for opening the container partially to regulate the release rate of the active ingredient. Figure 4 shows the effect of the opening of the mechanical container on the release rate from an air freshening gel contining rose.[29] A decrease of the opening from ¼ in. to ½ in. reduced the release rate by 25%.

Mechanical dispensers are usually very unique and ingenious devices and are granted design patents by the U.S. Patent and Trademark Office. A list of several such design patents is shown in Table 4.

E. Retrograde Chemical Reaction Systems

1. Reactive Hydrogels

Reactive hydrogels (Ionogels) are high-molecular-weight, crosslinked, hydrophilic polymers characterized by their ability to swell and soften in water without dissolving, as do the conventional hydrogels, but, at the same time, they can react with a wide range of substances. While conventional hydrogels are nonionic, inert, and nonreactive, ionogels can be made to contain different percentages of highly reactive cationic, anionic, amphoteric, or chelating co-monomers; therefore, they can be further reacted or combined with drugs, hormones, enzymes, metals, polymers, chemicals, and pesticides, provided that these contain reactive group.[30]

Anionic ionogels, containing acidic groups in their molecular structures, have been prepared based on heterocyclic monomers, such as N-vinyl lactam and hydroxyalkyl acrylate or methacrylate.[31,32] The polymerizable acidic monomer could be acrylic or methacrylic acid, sulfo-ethyl methacrylate or a sulfated or phosphated derivative of a hydroxyalkyl acrylate or methacrylate. These anionic ionogels may be combined by reaction or complexation with water-soluble or dispersible materials having an opposite charge, such as basic or cationic chemicals. Examples of agricultural chemicals which may be combined with the anionic ionogels and which are slowly released under conditions of use are the herbicides, atrazine, 2,4-dichloro-6-(o-chloroanilino)-s-triazine, and 2-chloro-4-ethylamino-6-isopropylamino-s-triazine. Examples of pharmaceutical agents include atropine, atropine-N-oxide dextroamphetamine, procaine, strychnine, basic narcotics such as codeine and morphine, anticonvulsants such as diphenyl hydantoin and pro-diphenyl hydantoin, antibiotics such as streptomycin, tetracycline, terramycine, and aureomycine, basic hormones such as insulin and thyroxin, and several vitamins, antihistamines, tranquilizers, and narcotic antagonists.

Cationic ionogels, containing basic (cationic) groups in their molecular structure, have also been prepared based on heterocyclic monomers, such as N-vinyl lactam and hydroxyalkyl acrylate or methacrylate.[33,34] The cationic monomer is an ester of acrylic or methacrylic acid with an amino alcohol, the terminal amino group of which may be quaternized, such as dimethylaminoethyl methacrylate or acrylate and their quaternized derivatives. These cationic ionogels are useful for combining, by reaction or complexing, with water-soluble or dispersible materials having an opposite charge, such as acidic or anionic chemicals. Examples of anionic agricultural chemicals which may be combined with the cationic ionogels and which are slowly released under conditions of use include 2,4-dichlorophenoxy acetic acid, 2,4,5-trichlorophenoxy acetic acid, and 2,2-dichloropropionic acid. Examples of pharmaceutical products include ascorbic acid, aspirin, barbiturates such as phenobarbital, pentobarbital, and barbital, penicillin such as penicillin G,K,X, and F, as well as such bactericides as hexachlorophene, halogenated phenols, and tribromosalicylanilide.

2. Hydrogels Containing a Biological Binding Agent

In this method, the biological binding agent, which contains bonding sites complementary to the bonding sites of the active chemical, is blended into a matrix of a water-insoluble, but water-swellable, hydrophilic polymer, such as a hydrogel. The molecular weight of the biological agent is large enough to cause its retention within the matrix and prevent its egress. The biologically active agent is then reacted with the binding agent to form the CR formulation. Alternatively, the reaction between the binding agent and the biologically active ingredient can take place before incorporation into the hydrogel.

Hydrogels containing zein as the biological binding agent and the bacteriostats, paradichlorobenzene, salicylic acid, and cetylpyridinium chloride, have been prepared.[35] These formulations were shown to provide sustained release of the bacteriostats for prolonged periods.

III. SEARCH OF U.S. PATENT LITERATURE

The chemical section and the general and mechanical sections of the Official Gazettes of the U.S. Patent and Trademark Office for all of 1977 and 3 months of 1978 were searched for any new CR methods and applications. The only contributions from the general and mechanical section were the design patents of mechanical dispensers, which are presented in Table 4. Microencapsulation patents pertaining to paper for use in pressure-sensitive copying were not considered. Since the searching of the Gazette is still an art, several patents pertaining to controlled release may have been missed; thus, no claims for complete coverage are being made.

Patents found in the search were categorized according to their general method of controlled release, i.e., (1) reservoir systems with a rate-controlling membrane, (2) reservoir systems without a rate-controlling membrane, (3) monolithic systems and laminated structures, (4) other physical methods, and (5) retrograde chemical reaction systems. The patents are listed in Table 1 through 5, respectively. The tables include patent number, company, title, application description, active agent, and CR method. From the applications point of view, the following is of interest:

1. A large number of patents were issued pertaining to the release of medicinals to the eye, bladder, stomach, uterus, vagina, and skin. Alza Corporation is the leader in this field with a cross section of technologies including osmotic pressure devices, laminated structures, and porous and nonporous polymeric membranes.
2. Patents pertaining to time-temperature indicators were granted for determining the condition of foods, pharmaceuticals, and biological preparations after exposure to unknown and possibly deleterious conditions of temperature and time. Reservoir systems with and without rate controlling membranes were used.
3. Design patents pertaining to decorative mechanical dispensers used for secondary control of the release of fragrances, fumigants, and volatile insecticides were issued. Airwick Corporation and Globol-Werk were most active in this area.
4. Patents pertaining to the protection of man and his environment were granted; included were patents for the control of flying and crawling insects, fertilizer for house plants, insect repellents, and release of fragrances in the home. Monolithic devices, microcapsules, and impregnated fibrous materials were utilized.

From the technology point of view, the only new methods uncovered are the reactive hydrogels and hydrogels containing a biological binding agent, as discussed in *Reactive Hydrogels* and *Hydrogels Containing a Biological Binding Agent*. These patents are listed in Table 5.

TABLE 1

Reservoir Systems with a Rate-Controlling Membrane

U.S. Patent No.	Company	Title	Application	Active agent	CR method
3996007	Bio-Medical Sciences, Inc.	Time-temperature Integrating Indicator	Indicates condition of foods, pharmaceuticals, and biological preparations	Acetic acid, ammonium carbonate, pH sensitive dyes	Polymeric membrane
4016251	Alza	Vaginal drug dispensing device	Controls fertility in the human	Progestational and estrogenic steroids	Polymeric membrane
4002458	3M	Controlled release capsules		Fertilizers and herbicides	Encapsulation
4056610	3M	Microcapsule Insecticide Composition	Controls flying and crawling insects	Pyrethroids	Microencapsulation
4014432	Lever Brothers	Product for treating fabric	Applies softeners and antistatic agents to fabrics during drying	Quaternary ammonium compounds	Porous membrane
4031202	Procter and Gamble	Controlled release contraceptive article	Releases medicinals to the body of animals and humans	Nonhormonal contraceptives	Porous membrane
4028876	L. Peska Associates	Apparatus for visually indicating elapsed time by a color change	Indicates condition of foods and pharmaceuticals	Oxalic acid and phenolphthalein	Polymeric membrane
4039653	DeFoney et al.	Long-acting articles for oral delivery and process	Medications for oral cavity	Eucalyptus oil, spearmint oil, and others	Erodible membrane
4040207	W. R. Lancaster	Self-fertilizing pot	A flower pot providing plant food continuously	Fertilizer	Microencapsulation

TABLE 1 (continued)

Reservoir Systems with a Rate-Controlling Membrane

U.S. Patent No.	Company	Title	Application	Active agent	CR method
4042336	Bio-Medical sciences, Inc.	Time-temperature Integrating Indicator	Indication of condition of foods and pharmaceuticals	Ammonium carbonate, trichoroacetic acid, maleic anhydride, and others	Polymeric membrane
4067961	Procter and Gamble Co.	Controlled release article	CR of nonhormonal contraceptives	Decyldimethylphosphine oxide and others	Microporous membrane
4069307	Alza	Drug delivery device comprising certain polymeric materials for controlled release of drugs	Release of medicine to eye, bladder, uterus, vagina, and skin	Pilocarpine, hydrocortisone, progesterone, etc.	Polymeric membrane
4079675	U.S. Army	Controlled solution releasing device	Desensitization of a mine and dispensation of fertilizer	Electrolytes and fertilizers	Porous membrane
4078423	Fuji Photo Film Co.	Pressure measuring sheet and method for pressure measurement using said sheet	Measurement of planar and linear pressures	Rhodamine B lactam and other dyes	Microencapsulation

| 4081188 | Wiggins Teape, Ltd | Paper having microcapsules deposited in depressions on a surface thereof has improved smudge-resistance characteristics | Paper for use in pressure-sensitive copying | Dyes | Microencapsulation |

TABLE 2

Reservoir Systems without a Rate-Controlling Membrane

U.S. Patent No.	Company	Title	Application	Active agent	CR Method
4011061	3M	Articles providing sustained release and method of making	Application of fertilizer and other agrichemicals	Urea and others	Water imbibed
4027420	U.S. Army	Air-dropped bait dispensers for attracting and killing the cotton boll weevil	Attracting and killing boll weevils	Grandlure and DDVP	Impregnated filter
4037352	Airwick	Electrical device which emits insecticidal vapors	Control of mosquitoes, flies, etc.	Pyrethrinoid insecticides	Impregnated fibrous material
4037353	Airwick	Device for emitting vapors of active substance	Control of mosquitoes, flies, etc.	Organophosphorus insecticides	Impregnated fibrous material

Table 2 (continued)
RESERVOIR SYSTEMS WITHOUT A RATE-CONTROLLING MEMBRANE

U.S. Patent No.	Company	Title	Application	Active agent	CR Method
4004685	Economics Laboratory	Treatment of fabric in machine dryer	Imparting properties to clothes during drying	Antistatic agent and fabric softener	Impregnated cloth
4023532	R. Goodwin	Face fly device	Repelling face flies from cows and other animals	Insecticides and repellents	Fabric wick
4021941	J. Caggiano	Disposable footwear	Application of medicaments to foot, e.g., treatment of athlete's foot		Absorbent material
4047505	Canada	Insect repellent collar	Collar for humans to repel biting insects	Deet, benzyl benzoate, and others	Open-cell foamed resin material
4058607	Airwick	Insecticide evaporator comprising a stabilizer	Control of flying insects	DDVP	Fibrous felt
4044707	General Foods Corp.	Interruptible time-temperature indicator	Indication of condition of foods	Methylene Blue	Filter Paper wick
4079547	F. Walker	Cellular moist film plant culture system	System for providing nutrients to growing plant	Limestone, calcium nitrate, and others	Sheets of porous material
4079700	W. S. Eshnaur	Dust bag for animal insecticide	Application of powdered insecticides to animals	Insecticides	Porous fabric

TABLE 3

Monolithic Systems and Laminated Structures

U.S. Patent No.	Company	Title	Application	Active agent	CR Method
3996348	A. H. Robbins	Insect-combating device	Control of houseflies, gnats, and mosquitoes	Naled insecticide	Monolithic polymer
3996933	Morton Gutnick	Intrauterine contraceptive devices and processes	A contraceptive device also useful for curing gonorrhea, syphillis, etc.		Monolithic elastomer
3996934	Alza	Medical bandage	Administration of drugs through skin	Nitroglycerin and others	Laminated structure
4008351	Sumitoma Bakelite	Film or sheet material having antibacterial antifungal activities	Wall coverings, ceiling materials, floorings, and synthetic paper	2-(4-Thiazolyl-benzimidazole and N-(fluorodichloromethyl-thio)-phthalimide	Monolithic polymer
4012221	International Copper Research Association	Slow release copper toxicant composition	Destroying trematode hosts	Copper ions	Monolithic elastomer
4012497	Schering	Drug excipient of silicone rubber	Prosthetic device and intrauterine device	Steroids and others	Monolithic elastomer
4014335	Alza	Ocular drug delivery device	Treatment of eye	Pilocarpine	Laminated structure
4019889	Swift Chemical	Slow-acting	Fertilizer for home and garden	Phosphorous, nitrogen, and potassium	Laminated structure fertilizer
4031894	Alza	Bandage to transdermally administer scopolamine to prevent nausea	Medicated bandage	Scopolamine	Laminated structure

TABLE 3 (continued)

Monolithic Systems and Laminated Structures

U.S. Patent No.	Company	Title	Application	Active agent	CR Method
4028045	W. Reiher	Specialized candle	Candle	Fragrances and insect repellents	Monolithic wax matrix
4035479	Celanese	Delayed and sustained urea release compositions	Ruminant feed supplement	Urea	Erodible (water soluble)
4043073	M. J. Basile	Method of treating soil for controlling termites and the like	Termite control	Mirex	Erodible matrix
4066754	Ralston Purina	Slow release bolus	Delivery of veterinary medicament in ruminant animals	Sulfamethazine and sulfathiazole	Erodible matrix
4056612	Stauffer Chemical	Air freshener gels	Air freshening air space	Rose, lemon, and others	Gel
4081273	Xerox	Migration imaging method	A migration imaging system	Selenium	Laminated structure
4077324	Xerox	Method of fountainless lithography	Printing without an aqueous foundation solution	Ink	Laminated structure
4077325	Xerox	Process for preparing waterless printing masters	Waterless lithographic printing	Ink	Laminated structure
4057029	Infratab	Time-temperature indicator	To indicate if product has been exposed to undesirable conditions	Lauryl diethylamine/ alizarin and others	Laminated structure

U.S. Patent No.	Company	Title	Application	Active agent	CR Method
4060084	Alza	Method and therapeutic system for providing chemotherapy transdermally	Provides transdermal chemotherapy	Scopolamine	Laminated porous structure

TABLE 4

Other Physical Methods

U.S. Patent No.	Company	Title	Application	Active agent	CR Method
3995631	Alza	Osmotic dispenser with means for dispensing active agent responsive to osmotic gradient	Release of medicine to eye, bladder, uterus, stomach, vagina, etc.	Proteins and other	Osmotic device
3995632	Alza	Osmotic dispenser	Release of pharmaceuticals to the body	Proteins and other	Osmotic device
4014334	Alza	Laminated osmotic system for dispensing beneficial agent	Oral and eye delivery of medicinals	Pilocarpine, methazolamide, and others	Osmotic device
4020156	Norda, Inc.	Controlled fragrance releasing crystal beads	Perfume sachets, bath beads, and control of malodors	Rose, perfume, and others	Water soluble carrier
D244775	The Puro Co.	Air freshener container	Release of fragrance		Mechanical dispenser
D224692	Globol-Werk	Air freshener or similar article	Release of fragrance		Mechanical dispenser

TABLE 4 (continued)

Other Physical Methods

U.S. Patent No.	Company	Title	Application	Active agent	CR Method
4034758	Alza	Osmotic therapeutic system for administering medicament	Drug release to uterus, eye, and stomach	Potassium chloride and others	Osmotic device
4036227	Alza	Osmotic releasing device having a plurality of release rate patterns	Drug delivery to eye, uterus, and stomach	Pilocarpine nitrate and Potassium chloride	Osmotic device
4036228	Alza	Osmotic dispenser with gas generating means	Oral effervescent antacids and analgesics	Acetylsalicylic acid, potassium chloride, and sodium bicarbonate	Osmotic device
D244200	American Home Products	Air freshener container or the like	Delivery of fragrances to air space	Fragrances and perfumes	Mechanical dispenser
D245705	Eikosha Co.	Fragrance emitter housing	Delivery of fragrances to air space	Fragrances	Mechanical dispenser
D245787	Shell Oil	Container for disseminating insecticidal or other vapors	Delivery of insecticidal vapors	Volatile insecticides	Mechanical dispenser
D245862	A. Bulloch	Container for solid vaporizer	Delivery of vapors to air space		Mechanical dispenser
D246052	Taisho Pharmaceutical	Fumigator	Delivery of fumigants to air space	Fumigants	Mechanical dispenser

Patent No.	Company	Device	Application	Active agent	Device type
D246319 D246317 D246110 D246316 D246318	Airwick	Dispenser for air treating material	Delivery of fragrances to air space	Fragrances	Mechanical dispenser
D246185	F. Brandburne	Air freshener or the like	Delivery of fragrances to air space	Fragrances	Mechanical dispenser
4058122	Alza	Osmotic system with laminated wall formed of different materials	Delivery of medicinals orally	Ascorbic acid, diazepan, and others	Osmotic device
4062649	Shell Oil	Depletion indicator for controlled release pesticide formulations	Determination of depletion of DDVP strips	DDVP, sodium hydroxide, and methyl red	Monolithic device and filter paper
4067692	R. W. Farris	Odor control device	Odor control device for heating and/or air conditioning system	Fragrances	Mechanical dispenser
D247249	Globol-Werk	Combined holder and mounting bracket for solid deodorant, perfume, insect repellent, or the like	Control the release of fragrances and repellents	Fragrances and repellents	Mechanical dispenser
4077407	Alza	Osmotic devices having composite walls	Delivery of medicaments orally	Potassium chloride, ascorbic acid, and others	Osmotic device
4038873	Big Three Industries, Inc.	Temperature monitor and indicator	Indication of condition of food after many days	Diazo and anthraquinnone dyes	Diffusion through liquid impregnated on paper carrier

TABLE 5

Chemical Reaction Systems

U.S. Patent No.	Company	Title	Application	Active agent	CR Method
4007258	Union Corporation	Sustained release pesticidal composition	Bactericidal applications	Cetylpyridinium chloride	Pesticide polymer in hydrogel
4058491	Plastomedical Sciences	Anionic hydrogels based on heterocyclic n-vinyl monomers	Release of agricultural and pharmaceutical agents	Atrazine, atropine, homatropine, and others	Reactive hydrogel
4060678	Plastomedical Sciences	Cationic hydrogels based on hydroxyalkyl acrylates and methacrylates	Release of agricultural and pharmaceutical agents	2,2-Dichloropropionic acid, ascorbic acid, aspirin, penicillin, etc.	Reactive hydrogel
4071508	Plastomedical Sciences	Anionic hydrogels based on hydroxyalkyl acrylates and methacrylates	Release of agricultural and pharmaceutical agents	Atrazine, atropine, homatropine, and others	Reactive hydrogel
4058491	Plastomedical Sciences	Cationic hydrogels based on heterocyclic n-vinyl monomers	Release of agricultural and pharmaceutical agents	2,2-Dichloropropionic acid, ascorbic acid, aspirin, penicillin, etc.	Reactive hydrogel

REFERENCES

1. Reyes, Z., Process for Making Microcapsules, U.S. Patent 3,173,878, 1965.
2. Herbig, J., Process for Making Capsules, U.S. Patent 3,161,602, 1964.
3. Macaulay, N., Recording Paper Coated with Microscopic Capsules of Coloring Material; Capsules and Method of Making, U.S. Patent 3,016,308, 1962.
4. Belcher, D. W., Smith, D. A., and Cook, E. M., Spray drying equipment in, *The Encyclopedia of Chemical Process Equipment*, Mead, W. J., Ed., Krieger Publishing, Huntington, N.Y., 1964, 861.
5. Scher, H. B., Great Britain Patent 1,371,179, 1975.
6. Scher, H. G., Microencapsulated pesticides, in *Controlled Release Pesticides*, Scher, H. B., Ed., ACS Symp. Series 53, American Chemical Society, Washington, D.C., 1977, 126.
7. Brynko, C., Oil Containing Capsules and Method of Making Them, U.S. Patent 2,969,330, 1961.
8. Berger, B. B., Electrostatic Encapsulation, U.S. Patent 3,208,951, 1965.
9. Bakan, J. A., Conrolled release pesticides via microencapsulation, in *Proc. 1976 Controlled Release Pesticide Symp.*, Cardarelli, N., Ed., University of Akron, Ohio, 1976, 1.1.
10. Langer, G. and Yamate, G., Apparatus for Electrostatic Encapsulation, U.S. Patent 3,294,704, 1966.
11. Langer, G. and Yamate, G., Apparatus for Electrostatic Encapsulation, U.S. Patent 3,159,874, 1964.
12. Barclay, W H., Mixing equipment in, *The Encyclopedia of Chemical Process Equipment*, Mead, W. J., Ed., Krieger Publishing, Huntington, N.Y., 1964, 658.
13. Akkapeddi, M. K., Halpern, B. D., Davis, R. H., and Balin, H., Cervical Hydrogel Dilator: a new delivery system for prostaglandins, in *Controlled Release of Biologically Active Agents*, Tanquary, A. C. and Lacey, R. E., Ed., Plenum Press, New York, 1974, 165.
14. Anon., Slow release achieved for macromolecules, *Chem. Eng. News*, 55, 26, 1977.
15. Tollar, M., Surgical suture materials coated with a layer of hydrophilic hydron gel, *J. Biomed. Mater. Res.*, 3, 305, 1969.
16. Drobnik, J., Spacek, P., Wichterle, O., Diffusion of antitumor drugs through membranes from hydrophilic methylacrylate gels, *J. Biomed. Mater. Res.*, 8, 45, 1974.
17. Molday, R. S. and Dreyer, W. J., Latex spheres as markers for studies of cell surface receptors by scanning electron microscopy, *Nature (London)*, 249, 81, 1974.
18. Steckler, R., U.S. Patent 3,532,679, 1970.
19. Steckler, R., U.S. Patent 3,878,175, 1975.
20. Jhon, M. S. and Andrade, J. D., Water and hydrogels, *J. Biomed. Mater. Res.*, 7, 509, 1973.
21. Lee, N., et al., Division Polymer Chemistry, American Chemical Society, Prepr. 15, 706, Washington, D.C., 1974.
22. Maron, S. H. and Prutton, C. F., *Principles of Physical Chemistry*, MacMillan, New York, 1965, 856.
23. Friedman, R. H., Krause, J. D., and Bradford, W. R., U.S. Patent 3,749,174, 1973.
24. Sayce, J. G. and Brown, D. J., U.S. Patent 3,969,280, 1976.
25. Chii-Fa, Lin, U.S. Patent 4,056,612, 1977.
26. Kydonieus, A. F. and Smith, I. K., unpublished data.
27. Nysted, L. N., U.S. Patent 3,745,213, 1973.
28. Olton, G. S., A new controlled-release insecticide for cockroach control, in *Proc. Int. Controlled Release Pesticide Symp.*, Harris, F. W., Ed., Wright State University, Dayton, Ohio, 1975, 258.
29. Kydonieus, A. F. and Smith, I. K., unpublished data.
30. Anon., Hydrogels and Ionogels, Report of Plastomedical Sciences, Inc., Briarcliff Manor, New York.
31. Steckler, R., U.S. Patent 4,036,788, 1977.
32. Steckler, R., U.S. Patent, 4,071,508, 1978.
33. Steckler, R., U.S. Patent 4,058,491, 1977.
34. Steckler, R., U.S. Patent 4,060,678, 1977.
35. Cohen, A. I., U.S. Patent 4,007,258, 1977.

INDEX

A

B

F

M